Natural Computing Series

More information about this series at http://www.springer.com/series/4190

Nelishia Pillay • Rong Qu

Hyper-Heuristics: Theory and Applications

Nelishia Pillay
School of Mathematics, Statistics
and Computer Science
University of KwaZulu-Natal
Pietermaritzburg, KwaZulu-Natal, South Africa

Rong Qu
School of Computer Science
University of Nottingham
Nottingham, UK

ISSN 1619-7127
Natural Computing Series
ISBN 978-3-030-07206-3 ISBN 978-3-319-96514-7 (eBook)
https://doi.org/10.1007/978-3-319-96514-7

Softcover re-print of the Hardcover 1st edition 2018

This Springer imprint is published by the registered company Springer Nature Switzerland AG
The registered company address is: Gewerbestrasse 11, 6330 Cham, Switzerland

To my parents Perumal Manickum and Sinathra Manickum
Nelishia Pillay

To my dear daughter Jessica Xue, who makes me strong and happy
Rong Qu

Foreword

It is hard to believe that Hyper-Heuristics have been part of my life for almost twenty years. The first paper I published on this topic was in 2001, together with my first PhD student (Eric Soubeiga) and his co-supervisor (Peter Cowling). Since that time I have published many other papers, on various topics, but my four most highly cited papers are on hyper-heuristics. It is gratifying to see the wider community also carrying out research in this area. A quick look at Scopus (27 Jun 2018) shows that there have been 729 papers published on the topic of hyper-heuristics since 2001, attracting 6,345 citations.

This book is an important text. It provides a single reference for anybody who has an interest in hyper-heuristics. The field has been missing a go-to book on hyper-heuristics for many years. This book fills that niche. Nelishia and Rong have produced a book that should be on the bookshelf of anybody who has an interest in this area.

The book can be read in its entirety, but it can also be dipped into as the need arises, choosing between the three distinct parts.

The first five chapters provide an excellent overview for those who are new to this area, or those who have focused on one aspect of hyper-heuristics but now wish to expand their horizons. Chapter 6 provides a more theoretical treatment.

Part II of the book should be of interest to the industrial community, as well as to the scientific community. The focus is on the application of hyper-heuristics to specific problem types. These include vehicle routing, nurse rostering, packing problems and examination timetabling. The presented domains are interesting in their own right but there is benefit in being able to generalize those domains to understand how they can be adapted for other problem types. Indeed, this is one of the biggest advantages of hyper-heuristics, more so than many other search methodologies. Hyper-heuristics are designed to adapt to a changing environment and their ability to produce high quality solutions across many domains is well documented in the scientific literature.

The final section of the book considers the past, present and future of hyper-heuristics. Two useful appendices conclude the book. One looks at the software frameworks that can assist when developing hyper-heuristics. The final appendix presents various benchmarks that are commonly used to evaluate hyper-heuristics.

I was honoured when Nelishia and Rong asked me to write this foreword and I highly recommend this book to those who are working with hyper-heuristics, or those who just want to know more about this exciting research area. This book will be useful to those who would like a gentle introduction to hyper-heuristics, as well those who have more experience in the domain. It will be the first place to look, when starting a new project in this area.

I hope that you enjoy reading this book as much as I did and I congratulate Nelishia and Rong on producing such an excellent tome.

Kuala Lumpur, Malaysia, June 2018 *Graham Kendall*

Preface

Hyper-heuristics is a fairly recent technique that aims at effectively solving various real-world optimization problems. This is the first book on hyper-heuristics, and aims to bring together both the theory and applications of hyper-heuristics, providing a solid foundation for the field.

The book is divided into three parts. The first part Hyper-Heuristics: Fundamentals and Theory first provides an overview of the four types of hyper-heuristics, namely, selection constructive, selection perturbative, generation constructive and generation perturbative hyper-heuristics in Chapters 2 to 5, respectively. Since the inception of hyper-heuristics, not much attention has been paid to the theoretical aspects of search in the search space of a heuristic. Chapter 6 focuses on this: here a formal definition of hyper-heuristics is provided, and a two-level framework defining the relationship between the heuristic and solution spaces is presented.

The second part of the book, Applications of Hyper-Heuristics, highlights the application of hyper-heuristics to solve problems arising in the real world, particularly in industry. The use of the different types of hyper-heuristics for the vehicle routing, nurse rostering, packing and examination timetabling problems is examined in Chapters 7 to 10, respectively. Research on cross-domain hyper-heuristics aims to increase the level of generality of hyper-heuristics by providing solutions across different problem domains instead of for a specific domain. Chapter 11 introduces cross-domain hyper-heuristics and highlights the advances made in this area.

Part three of the book, Past, Present and Future, first presents advanced topics in the field of hyper-heuristics, namely, hybrid hyper-heuristics, hyper-heuristics for automated design, automated design of hyper-heuristics, and hyper-heuristics for continuous optimization. A summary of the field and future research directions in hyper-heuristics are then provided.

In Appendix A, the book presents details of a hyper-heuristic framework and a toolkit that can be used to get started with research in hyper-heuristics without having to develop code from scratch. HyFlex is a framework for implementing selection perturbative hyper-heuristics to solve problems across domains, and provides libraries for six problem domains, with perturbative heuristics for these domains. EvoHyp is a toolkit comprising libraries for genetic algorithm selection hyper-heuristics and genetic programming generation constructive hyper-heuristics. Various publicly available benchmark sets are commonly used in the applications of hyper-heuristics presented in Chapters 7 to 10. Appendix B provides details of these benchmark sets.

Links to the websites for HyFlex and EvoHyp, the details of the benchmark sets and resources for the book can be found at https://sites.google.com/view/hyper-heuristicstheoryandapps

The book is aimed at postgraduate students, researchers and practitioners working in the field of hyper-heuristics, and can also serve as a textbook for postgraduate courses on hyper-heuristics.

Hyper-heuristics is a rapidly developing field with scope for growth, in terms of both applications and research. We hope that the book provides you with a foundation to get going with your research and applications in the field, contributing to further developments of hyper-heuristics. We enjoyed very much authoring the book, while overviewing the existing literature and establishing a bridge between applications and theory in hyper-heuristics. We hope you enjoy reading the book, and would welcome your advice and comments to us.

Pretoria, South Africa *Nelishia Pillay*
Nottingham, UK *Rong Qu*

May 2018

Acknowledgements

The authors would like to thank Prof. Graham Kendall for his invaluable feedback on the final draft of the book. We would also like to thank Prof. Kendall for his contribution in writing the foreword. Thank you to Mr. Derrick Beckedahl for developing the website for the book and EvoHyp and checking the final draft of the book. Last but not least a big thank you to Mr. Ronan Nugent for his advice and support throughout the writing of the book.

Contents

Acronyms and Notations

HH	Hyper-heuristic
P	The high-level combinatorial optimization problem, whose decision variables are heuristic configurations h
H	The high-level search space of heuristic configurations h for P
h	The heuristic configurations composed of low-level heuristics in L, $h \in H$
F	The high-level objective function for P, $F(h) \rightarrow R$
L	The given set of domain specific low-level heuristics used to create heuristic configurations h
p	The optimization problem at hand under consideration
s	The direction solutions for p
S	The low-level solution space for p, $s \in S$
f	The low-level objective function for $p, f(s) \rightarrow R$
SCH	Selection constructive hyper-heuristic
SPH	Selection perturbative hyper-heuristic
GCH	Generation constructive hyper-heuristic
GPH	Generation perturbative hyper-heuristic
I	Problem instances for problem p
i	Single problem instance of p
A	Attributes for a problem p, e.g. number of students in examination timetabling

Part I
Hyper-Heuristics: Fundamentals and Theory

Chapter 1
Introduction to Hyper-Heuristics

1.1 Introduction

Research into solving combinatorial optimization problems such as timetabling, vehicle routing and rostering problems has involved deriving techniques that improve the results obtained by existing techniques for known benchmark sets. These benchmark sets are made publicly available for performance comparisons of different techniques in solving these problems. This research has revealed that while a technique may produce the best results for one or two problem instances, quite often it performs poorly on other problem instances.

The field of hyper-heuristics emerged as an attempt to provide more generalized solutions to combinatorial optimization problems by performing well over a set of problems, rather than deriving techniques that produce good results for just a few problem instances for the domain. Hyper-heuristics achieve this by working in the *heuristic space* rather than the *solution space* [35, 151]. As such, hyper-heuristics either select or generate *low-level heuristics*, which are used to solve the problem at hand. Different techniques such as case-based reasoning, local search and genetic programming are employed by the hyper-heuristic to select or generate low-level heuristics. Low-level heuristics are described in Section 1.2. Section 1.3 presents a classification of hyper-heuristics.

1.2 Low-Level Heuristics

Hyper-heuristics either select low-level heuristics to construct or improve a solution, or create new low-level heuristics. Low-level heuristics are categorized as *constructive* or *perturbative*. These heuristics are usually defined for a particular problem domain and hence are problem specific.

Constructive heuristics are usually used to create an initial solution to a problem. This initial solution serves as a starting point for optimization techniques such

N. Pillay, R. Qu, *Hyper-Heuristics: Theory and Applications*,
Natural Computing Series, https://doi.org/10.1007/978-3-319-96514-7_1

as tabu search or simulated annealing in solving the problem. For example, in the domain of examination timetabling, constructive heuristics are used to select the examination to schedule next based on a measure of the difficulty of scheduling it. The constructive heuristics used to solve examination timetabling problems include largest degree, largest weighted degree, largest colour degree, largest enrolment and saturation degree [153]. In the case of population-based methods such as genetic algorithms, the initial population of timetables is created using a low-level heuristic rather than being randomly created. An initial solution, i.e. element of the population, is created by sorting the examinations according to their heuristic value and allocating them in order to the timetable [144]. Similarly, in the domain of one-dimensional bin-packing problems, constructive heuristics are used to select which bin to place an item in. Examples of constructive heuristics for this domain include first-fit, best-fit, next-fit and worst-fit.

Perturbative heuristics are applied to improve an existing initial solution created either randomly or by using a constructive heuristic. Low-level perturbative heuristics make changes to the initial solution, and have the same effect as a move operator in local search used to explore the neighbourhood of a search point. The perturbation made is dependent on the problem domain. For example, in the case of examination timetabling, examples of perturbative heuristics include swapping examinations between timetable periods, swapping rows in the timetable, deallocating an examination and allocating an examination.

1.3 Classification of Hyper-Heuristics

Given that hyper-heuristics either select existing low-level heuristics or generate new low-level heuristics, and these heuristics can be constructive or perturbative, hyper-heuristics are classified as being selection constructive, selection perturbative, generation constructive or generation perturbative [30].

Selection constructive hyper-heuristics select a low-level heuristic to apply at each point of the solution construction. Techniques employed by hyper-heuristics to select the low-level construction heuristics include case-based reasoning, local search methods, population-based methods, adaptive methods and hybrid approaches. Chapter 2 examines selection constructive hyper-heuristics in detail.

Selection perturbative hyper-heuristics select low-level perturbative heuristics to apply at each point of solution improvement. These can perform single-point or multipoint search. In the former case, the hyper-heuristic comprises two components, one for *heuristic selection* to select a low-level perturbative heuristic, and a second for *move acceptance* to determine whether the move made by the selected low-level heuristic should be accepted or not. Various techniques are used for heuristic selection and move acceptance. Selection perturbative hyper-heuristics performing multipoint search to select low-level heuristics use population based methods such as evolutionary algorithms to explore the heuristic space. The population-based technique by its nature performs both heuristic selection and move acceptance, and hence the hyper-heuristic does not contain separate components for these functions. Selection perturbative hyper-heuristics are described in Chapter 3.

Generation constructive hyper-heuristics create new low-level constructive heuristics for the problem domain. The generated heuristic is used to create an initial solution, which is optimized further using other techniques. Genetic programming [96] has chiefly been used by hyper-heuristics to generate construction heuristics. The components of the low-level heuristic include existing low-level heuristics or components of these heuristics as well as problem characteristics. These components are combined using arithmetic operators and conditional operators such as *if-then-else*. The evolved heuristic can be disposable or reusable [30]. Disposable heuristics are used to solve a particular problem instance. Reusable heuristics are generated using one or more problem instances and can be applied to unseen instances, i.e. problem instances not used in the induction and generation of the heuristic. Generation constructive hyper-heuristics are addressed in Chapter 4.

Generation perturbative hyper-heuristics produce new low-level perturbative heuristics. The new low-level heuristics are created by combining existing low-level perturbative heuristics and acceptance criteria using conditional statements, usually *if-then-else* statements. Examples of conditions used include whether a solution is found and whether a local optimum is reached [124]; see Chapter 5 for more details.

Chapter 2
Selection Constructive Hyper-Heuristics

2.1 Introduction

Selection constructive hyper-heuristics select a low-level heuristic at each point in the construction of a solution to a combinatorial optimization problem. As discussed in Chapter 1, the purpose of low-level construction heuristics is to construct complete solutions, or initial solutions for optimization. Solving a problem begins at an initial state and goes through a number of different problem states until the final state or solution state is reached.

A selection constructive hyper-heuristic selects the low-level construction heuristic to apply to go from one problem state to the next. The low-level heuristics are problem dependent. A formal definition of selection constructive hyper-heuristics is provided in Definition 2.1.

Definition 2.1. Given a problem p and a set of low-level construction heuristics $L = \{L_0, L_1, ..., L_n\}$ for the problem domain, a selection constructive hyper-heuristic constructs a solution s for p by selecting and applying a low-level heuristic from L to change from one problem state s' to the next s'', beginning at the initial state and stopping at the solution state s.

The hyper-heuristic generally employs a high-level technique such as a meta-heuristic or case-based reasoning to select the low-level heuristics. The algorithm generally employed by a selection constructive hyper-heuristic to solve the problem is outlined in Algorithm 1.

An overview of the four categories of techniques, namely, case-based reasoning, local search methods, population-based methods and hybridization and adaptive methods, employed by selection constructive hyper-heuristics is presented in the following sub-sections.

N. Pillay, R. Qu, *Hyper-Heuristics: Theory and Applications*,
Natural Computing Series, https://doi.org/10.1007/978-3-319-96514-7_2

Algorithm 1 Selection constructive hyper-heuristic algorithm

1: **procedure** SELECTIONCONSTRUCTIVEHYPERHEURISTIC(p, L)
2: initialize solution s to be empty
3: **repeat**
4: use technique T to select a low-level construction heuristic L_i from L
5: apply L_i to extend the solution s
6: **until** solution s is completely constructed
7: return s
8: **end procedure**

2.2 Case-Based Reasoning

Case-based reasoning (CBR) generally solves new problems (new cases) based on solutions of previous similar problems stored as previously solved source cases in a case base [1]. Retrieval and matching algorithms are used to find a source case that best matches the new case. The corresponding solution to the previous similar case is either used or adapted for the new case, based on the assumption that similar problems have similar solutions.

Case-based reasoning was one of the first techniques used to implement selection constructive hyper-heuristics to solve combinatorial optimization problems. Algorithm 2 illustrates the process of implementing a CBR selection constructive hyper-heuristic.

Algorithm 2 Implementing a CBR selection constructive hyper-heuristic

1: Create an initial case base (Algorithm 3)
2: Define the similarity measure, which calculates the similarity between cases
3: Refine the features and weights used in the similarity measure by evaluating the case base on a training set
4: Refine the set of cases by evaluating the performance of the CBR system on a training set

An initial case base is first created (see Algorithm 3). Each source case in the case base comprises a *description of the problem state* and the *best-performing heuristic(s)* for the source case. Each source case is usually described in terms of *features of the problem*, although for some problems this can be in a complex form. For example, for the examination timetabling problem typical features include the number of hard constraints, the number of soft constraints, the number of examinations and the density of the conflict matrix, amongst others. A heuristic is stored for each source case description and used to construct a solution. In the study conducted in [37] the five best-performing heuristics are stored for each source case, in ascending order of the objective value.

The set of appropriate features to represent and the set of source cases are crucial to build an effective CBR system. As depicted in Algorithm 3 an initial set of features needs to be further refined to obtain a set of features with higher effectiveness at constructing solutions. The initial set of features usually include all the possible

Algorithm 3 Creating an initial case base

1: Select an initial set of features
2: Select weights w_i for each feature i
3: Choose a set of problem states of differing characteristics
4: Solve the problems using different construction heuristics
5: Store the problem states as cases represented by the problem features and corresponding best-performing construction heuristic(s)

characteristics of the problem states. In the study by Burke et al. [40], the features are categorized as simple, complex, or combinations of subsets of this initial set.

To find a solution to a new problem, a source case that is most similar to the new problem case at hand is retrieved, and the retrieved heuristic or sequence of heuristics are used to construct a solution to the problem. A *similarity measure* based on the case features determines the similarity between cases. One commonly used similarity measure is the *nearest-neighbour* [37, 40], as depicted in Equation 2.1:

$$S(SC,P) = \frac{1}{\sqrt{\sum_{i=0}^{j} w_i * (fsc_i - fp_i)^2 + 1}} \tag{2.1}$$

SC refers to the source case retrieved from the case base and *P* refers to the new problem case being solved, and the *ith* feature is weighted by w_i. The weighted sum of all *j* pairs of features, fsc_i and fp_i for the *ith* feature of *SC* and of *P*, respectively, is calculated as *S(SC,P)* to define the similarity between *SC* and *P*, a higher value indicating a greater similarity.

Based on an initial case base and similarity measure, features used to represent source cases in the case base (line 4 of Algorithm 2) need to be refined to improve the performance of the CBR system. First a set of training cases are labelled against the best heuristics obtained for them using optimization methods. Then weights of the features in the similarity measure are repeatedly adjusted according to the best heuristics labelled for these training cases. This process continues until the heuristics of the cases retrieved using the similarity measure match most of those specified in the training cases. This training process can also be seen as a combinatorial optimization problem, and local search has been used in [40] to search for the best combination of features and their weights.

To further refine source cases, only relevant and useful cases that contribute to high accuracy of recommending the best heuristics are retained in the case base. Various techniques can be used for such system training based on a set of training cases. In Burke et al. [37], the "Leave-One-Out" method, which iteratively tests the effect of removing one source case at a time, is used to obtain the highest accuracy of retrieving the best heuristics.

CBR-based selection constructive hyper-heuristics have been used to solve problems in the domain of educational timetabling, namely, university course timetabling and examination timetabling [37, 40, 135].

2.3 Local Search Methods

Local search methods usually work in a solution space, iteratively exploring improved neighbourhoods of an initial solution using move operators until no further improvement can be made, i.e. a local optimum is reached. These methods generally differ in how they escape from local optima [19]. Tabu search (TS) attempts to escape local optima by using short-term memory to prevent neighbours from being revisited for a set number of iterations [70]. The current solution is replaced with the best solution in its neighbor on each iteration until the termination criteria are met. Variable-neighborhood search (VNS) escapes local optima by switching between different neighbourhood operators [74]. VNS iterates between the processes of shaking, local search and move. On each iteration a neigbourhood of the current solution is randomly selected and local search is applied. If the solution produced by the local search is better than the current solution, it replaces the current solution in the VNS. Iterated local search applies local search to an initial solution until a local optimum is reached, at which point perturbation is performed to escape the local optimum. The local search is applied in the new area of the search space to which the perturbation has led the search [106].

When employed by a selection constructive hyper-heuristic, a local search searches the heuristic space rather than the solution space. The local search explores the neighbourhood of a heuristic combination, which is composed of low-level construction heuristics for the problem domain. The process of constructing a solution using a heuristic combination is outlined in Algorithm 4. Each heuristic is applied t times to create a solution. The lower limit for t is 1 and the upper limit is problem dependent. An example of a heuristic combination is $h_4h_2h_2h_3h_1h_4h_1$. A construction heuristic may be included more than once, at different positions, in the heuristic combination.

Algorithm 4 Constructing a solution using heuristic combination h

```
1: procedure CREATESOLUTION(h, t)
2:     initialize an empty solution s
3:     for i=1 do length(h)
4:         for j=1 do t
5:             apply the ith heuristic h_i in h to extend the solution s
6:         end for
7:     end for
8:     return s
9: end procedure
```

The algorithm generally employed by a local search to explore the heuristic space is illustrated in Algorithm 5. The search is applied to an initial heuristic combination, which is either randomly created or composed of one specific low-level heuristic [27], e.g. $h_3h_3h_3h_3h_3$. The move operator (line 4) changes one or more low-level heuristics in the heuristic combination. Each heuristic combination is used to create a solution to the problem. The objective value of the resulting solution s forms the input to the mechanism of the particular local search employed to determine whether H is accepted or not. For some problem domains the heuristic combinations may result in solutions that are not valid, such as infeasible timetables [27]. The combinations resulting in invalid solutions are stored to ensure these areas of the search space are not visited again.

Algorithm 5 Local search on the heuristic space

```
1: Create an initial heuristic combination h=h1 h2...hn
2: procedure SEARCH(h)
3:     repeat
4:         Change one or more heuristics hi in h
5:         Use h to construct a solution s to the problem
6:         Calculate the objective value f of s
7:         Apply criteria specific to the local search used to decide whether to accept h given f
8:     until Termination criteria are met
9: end procedure
```

Moves made in the heuristic space correspond to "moves" in the solution space. Research has shown that, in selection constructive hyper-heuristics, small moves in the heuristic space [38], e.g. changing one low-level heuristic in the heuristic combination, can result in large moves in the solution space, thereby enabling the search to move more quickly through larger regions in the solution space [38]. This will be discussed further in Chapter 6.

In selection constructive hyper-heuristics, tabu search [38] and variable-neighborhood search [28] have been used to explore the heuristic space for both the examination and course timetabling problems. It was found in [151] that variable-neighborhood search and iterated-local-search-based hyper-heuristics perform the best for the benchmark course and examination timetabling problems based on comparisons of four local search methods. The study also analysed the search within two search spaces, namely the heuristic and solution space; see more details in Chapters 6 and 10.

2.4 Population-Based Methods

Whereas local search methods move from one point in the search space to the next, population-based searches explore multiple points simultaneously. The population of solutions represents different points in the search space. Evolutionary algorithms have chiefly been used to explore the heuristic space in the literature. Algorithm 6 presents a generational evolutionary algorithm. The genetic operators are applied to the heuristic combinations and hence perform intensification and diversification in the heuristic space.

Each element of the population, i.e. chromosome, is a heuristic combination. The combination comprises low-level construction heuristics, with each heuristic

Algorithm 6 Generational evolutionary algorithm

1: Create an initial population
2: **repeat**
3: Evaluate the population
4: Select parents
5: Apply genetic operators to the parents to create offspring of the new generation
6: **until** Termination criteria are met

in the combination representing a heuristic selected by the selection constructive hyper-heuristic. The selection as such is performed by the genetic algorithm by the process of fitness evaluation, selection and regeneration. Each chromosome is applied to solve one or more problem instances, and the fitness is the objective value in the case of a single problem instance or a function of the objective values for the different problem instances. If one problem instance is used for evaluation the aim is to evolve a heuristic combination specific to the problem at hand and the heuristic combination is disposable. More than one instance is used for evaluation to evolve a reusable heuristic combination. In this case the problem instances are divided into training and testing sets. The training set is used to evolve the heuristic combination and the testing set is a set of unseen problems on which the evolved heuristic combination is tested.

The representation used by the population-based approach has an effect on the performance of the hyper-heuristic. The simplest chromosome representation is a single heuristic combination consisting of low-level heuristics of a particular type. For example in [139] each chromosome is composed of graph colouring heuristics, which are used to create an initial timetable. The study compares three chromosome representations, namely, fixed-length, variable-length and N-times. In the fixed length representation, chromosomes have a fixed length equal to the number of examinations to schedule and each heuristic is used to schedule one examination. The length of a variable length chromosome is randomly chosen to be between one and a preset maximum. If the number of examinations is larger than the chromosome length, the combination is applied again beginning with the first heuristic. Similarly, if the length of the chromosome is larger than the number of examinations the remaining heuristics are not applied. In the case of the N-times representation each gene in the chromosome comprises an integer number n and a heuristic h, where h is used to schedule n examinations. Each chromosome is composed of m genes in the format $n_1h_1n_2h_2...n_mh_m$. The integers n_i must add up to the number of examinations to be allocated. The evolutionary algorithm hyper-heuristic employing the variable-length representation was found to perform better than the fixed-length and N-times representations. An evolutionary algorithm hyper-heuristic combining all three representations was investigated and produced better results than the variable-length evolutionary algorithm hyper-heuristic. For all representations the heuristic combinations evolved are disposable.

A chromosome can also comprise more than one type of low-level heuristic. In the study presented in [140] for the one-dimensional bin-packing problem, each chromosome comprises two heuristic combinations, one comprising heuristics to select which bin to place the chosen item in, and one to choose the item to place. The evolutionary algorithm hyper-heuristic using this representation was found to perform better than an evolutionary algorithm heuristic with chromosomes consisting of only a single heuristic combination comprised of bin selection heuristics. The generated heuristic combinations were disposable.

Different low-level heuristics perform well for different problem instances. Furthermore, a different construction heuristic would be more effective for different states of the problem leading to the solution state. Thus, chromosomes combining problem characteristics or the current state of the problem and low-level heuristics have also been used. In the chromosome $c_1c_2...c_mh_1h_2...h_n$, the first m genes represent the characteristics or current state of the problem instances, and the remaining n genes the corresponding low-level heuristics. Using this representation, each chromosome represents a condition-action rule, with the condition being the problem characteristic or current state and the action the construction heuristic to use [162].

Learning classifier systems are steady-state genetic algorithms incorporating reinforcement learning for fitness evaluation, in which each chromosome is represented as condition-action pairs, representing a classifier [24]. In the study conducted by Ross et al. [162], a learning classifier system is employed to evolve a reusable heuristic combination to solve the one-dimensional bin-packing problem. Each chromosome is a condition-action rule comprising of genes representing the current state of the problem and the corresponding heuristic to apply.

Messy genetic algorithms are a variation of genetic algorithms, in which chromosomes have variable length and solutions are evolved by combining shorter building blocks to produce solution chromosomes iteratively [71]. Ross et al. [160] use a messy genetic algorithm hyper-heuristic to solve the examination timetabling problem. The hyper-heuristic selects a construction heuristic to choose an examination and a period heuristic to select a period. In [161] a messy genetic algorithm is used to explore the heuristic space to solve the one-dimensional bin-packing problem. In this study a chromosome is composed of blocks. Each block is a condition-action rule with the condition presenting the problem state and the action the corresponding heuristic to apply. The evolved rules are reusable. The same approach is taken in [184] to solve the 2D-regular cutting-stock problem. Two types of heuristics are included in each block, namely, selection heuristics for selecting figures and objectives, and placement heuristics, which are used to place figures into objects. In both studies a training set is used to evolve a heuristic combination, which is applied to a test set. Each chromosome is applied to a different problem on each generation and the fitness of the chromosome is a function of its fitness over the generations and the number of problems the chromosome is applied to. Two crossover operators are used: one is applied to each block across parents and the other to the chromosome to exchange blocks. Mutation operators add and remove blocks from a chromosome, as well as change the gene in a block. The same approach is taken in [181] to solve the dynamic variable ordering in constraint satisfaction problems, and extended in [182] to solve the irregular packing problem. In this study the low-level heuristics are selection heuristics to choose the next variable in solving the problem.

In [189] a dispatching-rule based genetic algorithm (DRGA) is used to evolve rules to solve multi-objective and single-objective job shop problems. Each dispatching rule is composed of two sequences; the first is a sequence of low-level construction heuristics and the second a sequence of integers indicating the number of times each of the heuristics in the first sequence will be applied.

2.5 Hybridization and Adaptive Methods

Some of the earlier work in selection constructive hyper-heuristics did not employ a search to explore the heuristic space, but examined different hybridizations of low-level construction heuristics or adapted initial heuristic combinations in a second phase based on heuristic performance in the first phase. This section provides an overview of these hybrid and adaptive methods.

In the study in [41], the largest weighted degree and saturation degree construction heuristics are hybridized to create a restricted candidate list for GRASP for the examination timetabling problem. It was found that saturation degree is essential to find an initial feasible candidate solution, but it does not perform well in the early stages of timetable construction, as the majority of the examinations have the same saturation degree. However, largest weighted degree performs better in the early stages as more constrained examinations can be distinguished. In the hyper-heuristic, a switching point (when to start using the saturation degree heuristic) in the heuristic combination is thus adaptively determined until a feasible solution is found. In [151] a large number of random heuristic combinations of four widely used graph colouring construction heuristics are analysed to extract a trend in their application to construct solutions for examination timetabling and graph colouring problems. An adaptive hyper-heuristic is devised to hybridize the largest weighted degree heuristic into heuristic combinations.

Sabar et al. [165] combine and hybridize four low-level graph colouring construction heuristics for the examination timetabling problem. Four sequences of these heuristics are created. Each sequence is applied hierarchically: the second heuristic is used to break the tie from the use of the first heuristic and the third used if there is a tie for the first two construction heuristics in the sequence. The four heuristic sequences are used to create four lists of exams to be scheduled. The ranks of each examination in the four lists summed to provide an overall measure of the difficulty of scheduling the examination.

2.6 Discussion

The chapter overviews some of the methods employed by selection constructive hyper-heuristics. Case-based reasoning (CBR) is one of the first techniques used for this purpose. Such systems solve new problems based on knowledge of solving previous similar problems. The knowledge is extracted using off-line learning conducted on training cases with best constructive heuristics. The challenges posed by CBR include to determine how these constructive heuristics can be reused for similar problem states, and how to store such knowledge, i.e. the most appropriate features to describe and represent each case. More research findings based on a large number of training cases, and observations from other construction hyper-heuristics are needed to establish effective CBR systems.

Both local searches and evolutionary algorithms have been effective in exploring the heuristic space. Different research issues have been addressed. The set of low-level heuristics to use to compose heuristic combinations needs to be carefully chosen, and needs further research analysis. A large set with less useful constructive heuristics can lead to a search space too large to explore for an optimal heuristic combination within a limited runtime. In addition to low-level construction heuristics, the heuristic space may alternatively consist of condition-action rules, with the condition representing problem states and the action the corresponding heuristic to apply. It has also been shown that different low-level construction heuristics are needed at different points in constructing a solution, i.e. a different heuristic is needed for each problem state from the initial state to the solution state. Adaptive methods showed to be effective at adapting different types of constructive heuristics at different stages of solution construction.

In the case of population-based techniques such as evolutionary algorithms, the evolved heuristic combination or rule can be disposable or reusable. A disposable combination or rule is used to solve one problem instance. Reusable combinations or rules are evolved using a training set and are applied to unseen problems.

Table 2.1 provides a summary of application domains to which selection constructive hyper-heuristics have been applied and the techniques employed by the hyper-heuristics. The studies listed include early work in the field and domains for which there has been a fair amount of research into using selection constructive hyper-heuristics.

For all the domains, the selection constructive hyper-heuristics produced better results than each of the individual low-level heuristics. Results from the hyper-heuristics are also competitive with, and in some cases better on some problem instances than those of the state-of-the-art approaches. In some studies optimal results have been produced for a majority of the problem instances ([139, 161, 162, 184]), although this is not the main aim of selection constructive hyper-heuristics.

Table 2.1 Selection constructive hyper-heuristics for educational timetabling CBR: Case-based reasoning; LSM: Local search methods; EA: Evolutionary algorithms; HSM: Hybridization and adaptive methods

Problem	**CBR**	**LSM**	**EA**	**HSM**
Examination timetabling	Burke et al. [37, 40]	Burke et al. [27, 38], Qu et al. [28, 151]	Pillay [139] Ross et al. [160]	Burke et al. [41] Qu et al. [152] Sabar et al. [165]
Course timetabling	Burke et al. [37, 40] Petrovic et al. [135]	Burke et al. [38] Qu et al. [151]	-	-
One-dimensional bin-packing	-	-	Pillay [140] Ross et al. [161, 162]	-
Cutting stock problems	-		Terashima-Marin et al. [182, 184]	-
Dynamic variable ordering	-	-	Terashima-Marin et al. [181]	-
Job shop scheduling	-	-	Vazquez-Rodriguez and Petrovic [189]	- -
1D and 2D packing	- -	- -	López-Camacho et al. [105]	- -

Chapter 3
Selection Perturbative Hyper-Heuristics

3.1 Introduction

Selection perturbative hyper-heuristics select which low-level perturbative heuristic to apply at each point of improvement to a given initial complete solution to a problem. The initial solution is usually created either randomly or using a constructive low-level heuristic. It is usually iteratively refined by applying a perturbative low-level heuristic until there is no further improvement, measured using problem specific criteria such as the objective value of the perturbed solution. Starting from the initial problem state (solution), the application of each low-level perturbative heuristic results in moving from one problem state to the next until a final problem state, which cannot be improved further, is reached. A formal definition of selection perturbative hyper-heuristics is given in Definition 3.1.

Definition 3.1. Given a problem instance p, an initial solution s_0 and a set of low-level perturbative heuristics $L = \{L_0, L_1, ..., L_n\}$ for the problem domain, a selection perturbative hyper-heuristic *SPH* improves the solution s_0 by selecting and applying a perturbative heuristic L_i from L to get from one problem state s' to the next s'' until a problem state resulting in no further improvement of the solution s_i is reached.

As in the case of low-level constructive heuristics, the low-level perturbative heuristics are problem dependent. For example, in the case of solving the examination timetabling problem, a perturbative heuristic will swap the examinations of two timetable periods, while in the case of the travelling salesman problem a perturbative heuristic inserts a subset of cities at a new position in the route.

Selection perturbative hyper-heuristics employ *single-point* or *multipoint* search to select the low-level perturbative heuristics, as discussed further in Section 3.2 and Section 3.3, respectively. In single-point selection perturbative hyper-heuristics, two decisions are usually made, namely *heuristic selection* and *move acceptance* [30]. Multipoint selection perturbative hyper-heuristics employ population-based methods such as evolutionary algorithms to search the space of perturbative heuristics. By their nature, these search techniques perform both heuristic selection and

N. Pillay, R. Qu, *Hyper-Heuristics: Theory and Applications*,
Natural Computing Series, https://doi.org/10.1007/978-3-319-96514-7_3

move acceptance, and hence separate components for these functions are not needed [5, 45, 60, 73, 157].

3.2 Single-Point Search Selection Perturbative Hyper-Heuristics

Algorithm 7 depicts the general algorithm employed by single-point search selection perturbative hyper-heuristics.

Algorithm 7 Selection perturbative hyper-heuristic algorithm

```
1: procedure SELECTIONPERTURBATIVEHYPERHEURISTIC(p, L)
2:     create an initial solution s_0 using a random or constructive heuristic
3:     repeat
4:         use the heuristic selection technique h to select a perturbative heuristic L_i from L
5:         apply L_i to solution s_i to produce the perturbed solution s_{i+1}
6:         use the move acceptance technique M to accept the move or not
7:         if the move is accepted s_i=s_{i+1}
8:     until the termination criterion
9:     return s_i
10: end procedure
```

A termination criterion commonly used is that there is no further improvement in the solution s_i. Alternatively, the processes of heuristic selection and move acceptance can be performed for a set number of iterations. The following sections provide an overview of heuristic selection and move acceptance techniques.

3.2.1 Heuristic Selection Techniques

This section provides an overview of early and commonly used techniques for heuristic selection. It is by no means exhaustive as the list of techniques employed by selection perturbative techniques is rapidly growing.

The simplest heuristic selection technique is *random selection* [3, 99, 118], which randomly selects a perturbative heuristic from the available heuristics and applies it to the current solution s_i. A variation of the *random selection* technique is *random gradient* [3, 30], which selects a heuristic randomly and applies it iteratively, beginning with solution s_i, until there is no further improvement. *Random permutation* selects a sequence of perturbative heuristics randomly, they are applied in order [3, 30]. *Greedy* [30] applies all the perturbative heuristics in L and selects the heuristic producing the solution s_{i+1} with the best objective value.

Evolutionary algorithms have also been used for purposes of heuristic selection. Two methods used for heuristic selection are *tournament selection* and *fitness proportionate selection* [30]. These methods select a heuristic from the set L of available low-level perturbative heuristics. The fitness of each perturbative heuristic L_j is a problem specific measure, such as the objective value of the perturbed solution s_{i+1} resulting from applying L_j to s_i. In the case of tournament selection, a set of heuristics of fixed size is randomly selected from L, and the heuristic producing the solution with the best objective value is selected. On the other hand fitness proportionate selection creates a pool of heuristics based on the fitness of each heuristic L_i in L, and a heuristic is randomly selected from this pool.

The concept of a *choice function* was introduced for heuristic selection [30, 55, 89, 99]. A choice function calculates a rank for each heuristic L_j in L based on its performance, i.e. the improvements it has produced thus far and when it was last applied during the process. The heuristic with the best rank is selected and applied. The rank for each h_i is calculated using the following formulae [55, 89, 99]:

$$f(h_i) = \alpha f_1(h_i) + \beta f_2(h_i) + \delta f_3(h_i) \tag{3.1}$$

$$f_1(h_i) = \sum_n \alpha^{n-1} \frac{I_n(h_i)}{T_n(h_i)} \tag{3.2}$$

$$f_2(h_j, h_i) = \sum_n \beta^{n-1} \frac{I_n(h_j, h_i)}{T_n(h_j, h_i)} \tag{3.3}$$

$$f_3(h_i) = \tau(h_i) \tag{3.4}$$

f_1 in equation 3.2 is a measure of the recent performance of heuristic h_i over its previous n invocations. $I_n(h_i)$ is the change in objective value from the last invocation. Similarly, $T_n(h_i)$ is the difference in the time since the last invocation of the heuristic h_i. f_2 in equation 3.3 is a measure of the pairwise performance of h_i with all other heuristics h_j over n invocations of successive application of h_j and h_i. $I_n(h_j, h_i)$ is the difference in objective values from one successive application of h_j and h_i to the next. $T_n(h_j, h_i)$ is the difference in time since the last successive invocation of h_j and h_i. f_3 indicated in equation 3.4 is a measure of the time taken in CPU seconds since the heuristic was last applied during the improvement process. The parameters α, $\beta \in [0,1]$ set the importance of the recent performance of heuristic h_i. δ is a real-valued parameter used to maintain diversity. The basis of the choice function is reinforcement learning.

Reinforcement learning has also been successfully used for heuristic selection in selection perturbative hyper-heuristics [55, 85, 131, 132]. It assigns a score to each heuristic in the set L based on its performance during the improvement process. At the beginning of the improvement process all the heuristics are assigned the same score. During the process if a low-level heuristic L_i results in an improvement of a candidate solution its score is increased whereas if it results in a worse solution the score is decreased. The heuristic with the best score at the particular point of improvement is selected and applied.

A tabu list [70] has also been used as part of the heuristic selection component [36, 90] to prevent a poorly performing heuristic from being reused or to prevent the use of the same heuristic for a number of iterations in the improvement process.

Stochastic methods such as Markov chains are also used for heuristic selection [91]. Each chain is composed of low-level heuristics from L. The heuristics are not applied in sequences, a transition probability is associated with each heuristic and is used to decide which heuristic to apply next. Roulette wheel selection is used to choose the next heuristic to apply based on the transition probability.

Depending on the technique employed by the heuristic selection component, *learning* may or may not take place in selecting a heuristic. For example, in the case of random selection or random gradient, no learning is performed. However, reinforcement learning involves learning in terms of the performance of the low-level heuristics on previous iterations of the improvement process.

3.2.2 Move Acceptance Techniques

This section provides an account of early and commonly used move acceptance techniques. As in the case of heuristic selection techniques, the list of move acceptance methods successfully employed by selection perturbative hyper-heuristics is growing and the overview provided is not exhaustive. Move acceptance criteria are used to decide whether to accept the solution s_{i+1} produced by applying a perturbative heuristic L_j to s_i. The outcome of the move acceptance method is to either accept s_{i+1}, in which case it replaces s_i, or reject s_{i+1}, in which case s_i remains unchanged. The quality of solutions is measured in terms of the objective value of s_{i+1}.

The simplest move acceptance approach *accept all moves* [30] accepts the resulting solution produced by applying heuristic L_j irrespective of the quality of the perturbed solution s_{i+1} produced. A variation of this method accepts heuristics producing a worse perturbed solution s_{i+1} with a specified probability [91]. An extended technique *accept improving moves* [55, 99] only accepts the application of the heuristic L_j to s_i if there is an improvement in quality of the resulting solution s_{i+1}. Similarly, *accept equal and improving* [118, 131] accepts moves producing solutions of the same or better quality. Misir et al. [118] extend this idea and introduce two new move acceptance approaches, namely, Iteration Limited Threshold Accepting (ILTA) and Adapted Iteration Limited Threshold Accepting (AILTA). ILTA generally accepts improving and equal moves, and moves producing worse solutions if the current iteration exceeds the specified iteration limit and the fitness of the produced solution is less than a factor, specified by the threshold, of the best fitness obtained thus far. Both the iteration limit and the threshold value are parameters. AILTA is similar to ILTA but allows for the threshold value to be adapted if there is no improvement in the solutions perturbed.

Local search techniques have also been used for the purposes of move acceptance. These include simulated annealing [3, 55, 85], late-acceptance hill climbing

[55] and great deluge [3, 132]. This involves using the mechanism employed by the local search to accept a move to decide whether to accept the solution resulting from the application of the heuristic. The heuristic selected by the heuristic selection component is applied to solution s_i to produce the perturbed solution s_{i+1}. The acceptance mechanism of the local search is used to determine whether to accept or reject s_{i+1}. If s_{i+1} is an improvement over s_i it is accepted; otherwise the criterion specific to the local search for accepting worsening moves is applied. For example, in the case of simulated annealing a worse perturbed solution is accepted based on the temperature value.

3.3 Multipoint Search Selection Perturbative Hyper-Heuristics

Multipoint search selection perturbative hyper-heuristics employ population-based methods such as genetic algorithms [73, 156, 157], particle swarm optimization [5] and ant colonization [45, 60], to explore the heuristic space. The hyper-heuristic produces a heuristic [5] or a sequence of heuristics [5, 45, 73, 157] to improve an initial solution created randomly or using a constructive heuristic. When using particle swarm optimization each particle represents a heuristic or sequence of heuristics [5]. Similarly, in the study employing ant colonization [45] each ant chooses the next heuristic to apply. Genetic algorithms explore the space of low-level heuristic sequences [73, 156, 157]. In the case of a single heuristic this is applied to improve the initial solution, while a heuristic sequence is applied iteratively with each heuristic of the sequence applied in order to improve the initial solution. This process is depicted in Algorithm 8.

Algorithm 8 Applying a perturbative heuristic sequence

1: Given an initial solution s_0 and heuristic combination $h = h_1 ... h_n$
2: **for** $i \leftarrow 1, n$ **do**
3:     Apply the chosen h_i to s_i to create s_{i+1}
4: **end for**
5: Report s_n

Genetic algorithms have been the most popular multipoint search method employed by selection perturbative hyper-heuristics. Each element of the population, i.e. a chromosome, is a sequence of heuristics [73, 156, 157]. Research has shown that variable-length chromosomes are more effective than fixed-length chromosomes in genetic algorithm selection perturbative hyper-heuristics [73]. Each sequence is created by randomly selecting low-level perturbative heuristics from the available set of perturbative heuristics for the problem domain. An initial solution is created randomly or using a constructive heuristic and is used to calculate the fitness of each chromosome on each generation. The fitness of the chromosome is determined by applying it to the initial solution using Algorithm 8. The fitness of the chromosome is a function of the objective value of the resulting perturbed solution s_n.

3.4 Discussion

The majority of the research on selection perturbuative hyper-heuristics has focused on single-point search combining heuristic selection techniques and move acceptance criteria. The heuristic selection techniques used range from simple techniques with no learning that randomly select a perturbative heuristic, to approaches with learning and that select the heuristic based on its performance in previous iterations during the improvement process. The simplest move acceptance techniques are deterministic and accept all moves or improving and/or equal moves only. Variations include approaches that will also accept worsening moves based on certain criteria, e.g. a specified threshold value or the iteration of the improvement process. The acceptance mechanisms of local searches such as simulated annealing, great deluge and late-acceptance hill climbing have also proven to be effective for the purposes of move acceptance.

More recently multipoint search techniques such as genetic algorithms and particle swarm optimization have been employed by selection perturbative hyper-heuristics. An area that needs further investigation is comparative studies of single-point search and multipoint selection perturbative hyper-heuristics for different problem domains. Tables 3.1 and 3.2 list some of the application domains that single-point search and multipoint search hyper-heuristics have been applied to, respectively.

Table 3.1 Single-point search selection perturbative hyper-heuristic applications

Problem domain	Heuristic Selection	Move Acceptance
Maximum satisfiability [99]	Random, choice function	Improving only
Nurse rostering [36]	Reinforcement learning with tabu list	Accept all moves
University course timetabling [36]	Reinforcement learning with tabu list	Accept all
Multidimensional knapsack problem [36]	Random, choice function, reinforcement learning	Accept all, improving only, late acceptance, simulated annealing
Examination timetabling [90]	Tabu list	Accept all
Examination timetabling [132]	Reinforcement learning	Great deluge
Examination timetabling [131]	Reinforcement learning	Improving or equal
Home care scheduling [118]	Random	Improving or equal, iteration limited threshold accepting, adaptive iteration limited threshold accepting

As the field is developing, the range of techniques employed by selection perturbative hyper-heuristics is growing. Burke et al. [34] present a Monte Carlo selection perturbative hyper-heuristic framework that can be used with different approaches

Table 3.2 Multipoint Search Selection Perturbative Hyper-Heuristic Applications

Problem domain	**Multipoint Search Hyper-Heuristic**
Geographically distributed course timetabling problem	Genetic algorithm [73]
Student project presentation problem	Genetic algorithm [73]
Travelling tournament problem	Ant colonization [45]
Nurse rostering problem	Genetic algorithm [156]
School timetabling	Genetic algorithm [157]
Resource scheduling in a grid environment	Particle swarm optimization [5]
Set covering problem	Ant colonization [60]

for heuristic selection and move acceptance. Three Monte Carlo move acceptance approaches are used in this framework, namely, simulated annealing, simulated annealing with reheating and exponential Monte Carlo. In [91] Markov chains are used for heuristic selection. Misir et al. [121] have introduced the use of an automaton for heuristic selection.

Along with the research developments in meta-heuristics and evolutionary algorithms, a range of different perturbative move operators and acceptance criteria have been developed in various combinatorial optimization problems. The hybridizations of these with other techniques showed to be effective when adapted in selection perturbative hyper-heuristics. For example, in vehicle routing problems (see Chapter 7), both perturbative and constructive heuristics are selected by the high-level search to construct and improve solutions. In nurse rostering (see Chapter 8), low-level perturbative heuristics and acceptance criteria are selected as pairs at each decision point during the solution improvement process.

Chapter 4
Generation Constructive Hyper-Heuristics

4.1 Introduction

In solving combinatorial optimization problems, a low-level constructive heuristic is used to create an initial solution, which forms a starting point for optimization techniques to solve the problem. These heuristics are problem dependent and are rules of thumb, manually derived based on human intuition. Deriving constructive heuristics is a time-consuming process. Generation constructive hyper-heuristics aim to automate this process by generating low-level constructive heuristics using a given set of *problem attributes*. Automating this process reduces the man-hours involved in deriving low-level heuristics and may lead to the induction of new constructive heuristics that humans would not think of. This allows constructive heuristics to be tailored for a particular problem instance or to be induced for different *classes* of problems. Hence, the generated heuristic can be *disposable*, i.e. created for a specific problem instance, or *reusable*, i.e. used to solve similar unseen problems [30]. A formal definition of generation constructive hyper-heuristics is provided in Definition 4.1.

Definition 4.1. Given a problem instance i or a set of problem instances $I = \{I_0, I_1, ..., I_m\}$ and a set of problem attributes $A = \{A_0, A_1, ..., A_n\}$ for a problem domain, a generation constructive hyper-heuristic generates a new low-level constructive heuristic lch, using the attributes in A, to produce an initial solution for either i or the problems in I and similar problems.

The derived low-level heuristic is essentially a priority function that is used to order events or entities to be chosen to create a solution. As such the derived low-level heuristic is an arithmetic function or rule composed of the attributes and operators. Genetic programming [96] and variations thereof have chiefly been used for inducing these low-level heuristics. The hyper-heuristic achieves generality by using the same technique to derive heuristics for different problem instances and domains with the only change being the set of attribute values A used, which is problem dependent. However, the low-level heuristic induced may generalize or not, i.e. it can be *reusable* or *disposable*.

N. Pillay, R. Qu, *Hyper-Heuristics: Theory and Applications*,
Natural Computing Series, https://doi.org/10.1007/978-3-319-96514-7_4

4.2 Attributes and Representation of Low-Level Heuristics

Derived low-level heuristics comprise attributes of the problem and operators. Hence, methods used to create low-level heuristics combine or configure the attributes and operators in some way. It is important that an appropriate set of attributes is chosen and that all aspects of the problem domain are represented. However, including too many attributes will result in a larger heuristic space, which may lead to high processing times or a suitable heuristic not being found. According to Branke et al. [21], the attributes should be in their most basic form, and it should be left to the hyper-heuristic to create aggregated characteristics combining them. The attributes for a problem domain include:

- Characteristics of the problem - These are represented as variables in the induced heuristic, evaluated to a numerical value, for example, for the one-dimensional bin-packing problem, the capacity of the bin, fullness of the bin, and size of the item to be placed.
- Existing low-level constructive heuristics - The attributes can also include existing low-level heuristics that have been manually derived. For example, the largest degree, largest enrolment, largest weighted degree and saturation degree heuristics for the domain of examination and course timetabling.
- Components of existing low-level constructive heuristics - The basic components making up existing low-level constructive heuristics may be more representative of the problem domain than the heuristic as a whole. Existing low-level heuristics are decomposed into basic components, and used as attributes.

The attributes are configured into one of the two representations, namely, *arithmetic functions* or *rules*, to create new low-level heuristics. An arithmetic function is used to combine attributes with standard arithmetic operators, namely, addition, multiplication, subtraction and division. Constant values may also be included in the combination. The arithmetic function is used to calculate the priority of choosing events or entities when creating a solution. For example, in solving the examination timetabling problem the arithmetic function is used to calculate the difficulty of allocating examinations. The examinations are sorted in descending order according to this value, and allocated in this order to timetable periods to create an examination timetable. The arithmetic function may also contain a relational operator such as $\leq$, which returns a value of 1 if its first argument is less than or equal to its second argument, or -1 otherwise [32, 57]. Alternatively, the arithmetic function can be a weighted sum of attributes [21]. In this case the amount that each attribute contributes to the priority function is determined by the generated weight. The hyper-heuristic induces a combination of heuristics and weights, where the weights are constant integer or real values.

Rules comprise a condition component and an action component. The conditions include probabilities for probabilistic branching [11], how much of the solution has been created [11], or a comparison of attribute values, e.g. the number of enrolments for one examination is less than or equal to that for another examination [137]. Actions vary from existing low-level heuristics or components of low-level heuristics

[11], arithmetic expressions combining the problem characteristics, or the entity which should be given priority, e.g. which of two examinations should be scheduled next for the examination timetabling problem [137]. In the domain of production scheduling the generated heuristics are dispatching rules [21]. The dispatching rule produces a priority index for each job, which is a combination of problem characteristics and arithmetic operators.

Genetic programming and variations of genetic programming, such as grammar-based genetic programming [110] and grammatical evolution [129], have been employed by generation constructive hyper-heuristics to generate low-level constructive heuristics. In the case of grammar-based genetic programming and grammatical evolution, a grammar is used to define the structure of the arithmetic function or rule to represent the low-level heuristic. This also ensures that the functions and rules induced have feasible syntax. It also reduces the search space by restricting the search to areas with feasible functions and rules.

4.3 Genetic Programming

Genetic programming is an evolutionary algorithm that explores a program space rather than a solution space [96]. Programs can represent arithmetic functions or algorithms, which, when executed, will produce a solution to the problem at hand. Each program is represented as an expression tree. The generational genetic programming algorithm is depicted in Algorithm 9.

Algorithm 9 Genetic programming algorithm

1: Create an initial population
2: **repeat**
3: Evaluate the population
4: Select parents
5: Apply genetic operators to the parents to create offspring of the new generation
6: **until** Termination criteria are met

The algorithm begins with an initial population of programs, each an expression tree representing a new constructive heuristic. A fitness function is applied to evaluate each program in the population, i.e. how good it is at solving the problem at hand. In the case of evolving constructive heuristics the fitness of each expression tree is determined by the resulting solution created using the program tree. A selection method chooses parents based on their fitness to create offspring of successive generations. Tournament selection is generally used for genetic programming [96]. Genetic operators including reproduction, mutation and crossover are generally applied to the selected parents to create offspring of the next generation.

Each program in the population is created by randomly selecting elements from a function set and a terminal set until a maximum tree depth is reached. Elements of the function set are usually operators such as arithmetic operators and if-then-else statements, which form the internal nodes in the expression tree. Elements of the terminal set form leaf nodes in the expression tree, and act as arguments for the elements of the function set. To evolve constructive heuristics, the terminal set contains constants and variables representing the attributes of the problem. The function set comprises arithmetic operators in the case of arithmetic functions, and this set is extended to include an if or if-then-else statement when creating arithmetic rules.

In evolving constructive heuristics, strong typing is used to ensure that the trees produced by the initial population generation and the genetic operators represent valid heuristics. Elements of the function and terminal sets are assigned a type. A type is also specified for the arguments of function nodes. For example, an if-then-else statement will have a first argument of type Boolean and if the actions for the rule are arithmetic its remaining two children will be of type real. In some studies Boolean operators are treated as integer values, and typing is not needed as both functions and terminals will evaluate to a numerical value.

Grammar-based genetic programming is a variation of genetic programming in which a grammar dictates the structure of the expression tree [110]. This ensures that the expression trees created in the initial population and by mutation and crossover are syntactically correct. This restricts the search to areas of the search space that contain valid expression trees, and also reduces the space to be searched. Grammatical evolution is another variation of genetic programming that aims to reduce redundant code in the evolved programs [129]. Binary chromosomes are converted to denary values which are in turn mapped to production rules of a grammar in Backus-Naur format. In both grammar-based genetic programming and grammatical evolution, the grammar specifies how the attributes should be combined with different operators, e.g. arithmetic operators, if-then-else statements, when evolving low-level constructive heuristics. The study conducted by Harris et al. [75] shows that the variation of genetic programming used by the hyper-heuristic can effect the performance of hyper-heuristic.

4.4 Disposability vs. Reusability

The generated heuristic can be disposable or reusable [30, 33]. In the case of disposable heuristics, the hyper-heuristic performs online learning to evolve a heuristic for a particular problem instance [11, 175]. The generated heuristic is tailored for the particular problem instance. As discussed in the previous section on genetic programming algorithms which induce a heuristic, the fitness of the population of candidate heuristics must be calculated. In the case of disposable heuristics, the fitness of a program tree in the population is determined by the objective value of the solution produced by the program or a function of the this objective value. For example, in the domain of examination timetabling, the fitness is a function of the cost of hard constraints and soft constraints in the constructed timetable [142].

A reusable heuristic is usually evolved for a class of similar problems [82], and can be used to create an initial solution for other problem instances as well. A potential heuristic is applied to a set of problem instances, namely, the training set, to calculate the heuristic's fitness. One of the challenges is choosing an appropriate set of training instances. As highlighted in [21], training on too few instances can lead to overfitting, thus the heuristic may not perform well for other problem instances in the class. However, including too many problem instances will consume excessive computational time. This is easier to determine in some domains than others, for example for the job shop scheduling problem domain the benchmark set is divided into subsets of problem instances according to the number of jobs and machines [111]. Each subset corresponds to a problem class and a different heuristic is evolved for each class. Some instances in a subset thus form the training set and the remainder the test set. The research conducted in [82] shows that for some domains it may not be possible to induce heuristics that are as effective on unseen problems, and better results are achieved with disposable heuristics.

In calculating the fitness of a candidate heuristic, the heuristic is used to solve all the problems in the training set, and the fitness function is a function of the objective values obtained for each problem instance in the training set. The simplest function used is to sum the objective values [32, 82]. Alternatively, the average objective values or the sum of the deviations of the objective values from the known optimum for each problem instance can be used as a fitness function [21]. In order to generate more general heuristics, the instances in the training set can be chosen from different classes of problems. In the study conducted by Burke et al. [33] this proved to be effective in inducing general heuristics that were effective over all classes of problems for the two-dimensional strip packing problem.

4.5 Discussion

The aim of generation constructive hyper-heuristics is to produce low-level constructive heuristics. These were previously derived manually based on human intuition, which is a time-consuming and laborious process [21]. Hence automating this process will remove the onus from researchers and practitioners. Research has shown that different low-level constructive heuristics are effective for different classes of problems, and for some problem domains it is more effective to generate disposable heuristics for each problem instance. Deriving low-level constructive heuristics then becomes expensive to do manually [56] and is probably an intractable task.

Hence, two criteria should be used in assessing the performance of generation constructive hyper-heuristics, namely, the time it takes to generate these heuristics and the performance of the generated heuristics compared to existing manually derived heuristics. The time taken by the generation constructive hyper-heuristic should be less than it takes to manually derive these heuristics [11, 33]. It cannot be expected that the performance of the generated low-level heuristics will be comparable to the state of the art [32, 56]. As with manually derived heuristics, the aim of these heuristics is to provide a starting point for optimization techniques. Thus, the automatically generated heuristics should perform at least as well as the manually derived heuristics. However, from the research conducted in this field thus far, the heuristics produced by generation constructive hyper-heuristics have showed to outperform the existing heuristics.

Another important consideration is the interpretability of the generated constructive heuristics. Is it necessary for the generated heuristic to be readable to determine what it is doing, or should the generation constructive hyper-heuristic perform as a black box? If the former is required, grammatical evolution a better option to evolve heuristics, as standard genetic programming is susceptible to the growth of redundant code called *introns*. This reduces the readability of the evolved heuristics. Genetic programming can be employed with a tree size limit on the evolved trees, with trees exceeding the limit penalized as part of the fitness [32].

There has been a fair amount of research into the use of generation constructive hyper-heuristics for inducing low-level constructive heuristics for combinatorial optimization problems. Table 4.1 provides a summary of the genetic programming variations for different combinatorial optimization problems. While genetic programming, grammar-based genetic programming and grammatical evolution have predominantly been used by generation constructive hyper-heuristics for heuristic induction, some studies have investigated other techniques for this purpose. In the study conducted by Sim and Hart [174] another variation of genetic programming, namely single-node genetic programming, is used to evolve low-level constructive heuristics for the one-dimensional bin-packing problem. In [133] a genetic algorithm is used to evolve low-level constructive heuristics for the online one-dimensional bin-packing problem. Each heuristic is a policy matrix indicating the weight for packing an item in a bin depending on the residual space of the bin. The item with the highest weight is packed in the bin.

Table 4.1 Generation constructive hyper-heuristics. GP: Genetic Programming; GBGP: Grammar-Based Genetic Programming; GE: Grammatical Evolution

Problem domain	GP	GBGP	GE	Other
Examination timetabling	Pillay [137]	Bader-El-Den et al. [11]	-	-
Course timetabling	Pillay [141]	-	-	-
School timetabling	Pillay [138]	-	-	-
One-dimensional bin packing	Burke et al. [32] Hyde [82]	-	-	Sim and Hart [174] Özcan and Parkes [133]
Two-dimensional bin packing	Burke et al. [33] Hyde [82]	-	-	-
Three-dimensional timetabling	Hyde [82]	-	-	-
Vehicle routing	Sim and Hart [175]	-	Drake et al. [57]	-
Multidimensional knapsack problem	Drake et al. [56]	-	-	-
Constraint satisfaction problems	-	Sosa-Ascencio et al. [178]	-	-
Production scheduling	Branke et al. [21]	Branke et al. [21]	-	-

Chapter 5
Generation Perturbative Hyper-Heuristics

5.1 Introduction

Low-level perturbative heuristics are used to improve a solution created either randomly or using a constructive heuristic for a combinatorial optimization problem. The low-level perturbative heuristics are problem dependent, and often move operators defined for the problem domain when solving the problem using local search techniques, e.g. the 2-opt move operator for the travelling salesman problem, are used as perturbative heuristics. Hence, these are also referred to as local search operators. Generation perturbative hyper-heuristics aim at creating new low-level perturbative heuristics for a problem domain or instance. These heuristics are created by combining or configuring existing low-level perturbative heuristics and/or components of these heuristics. Genetic programming and variations thereof, e.g. grammatical evolution, have chiefly been used to combine these heuristics, and components with conditional branching and iterative constructs, to create new heuristics.

A formal definition of generation perturbative hyper-heuristics is provided in Definition 5.1.

Definition 5.1. Given a problem instance i or a set of problem instances $I = \{I_0, I_1, ..., I_m\}$ and a set of low-level perturbative heuristics and/or components of heuristics $C = \{C_0, C_1, ..., C_n\}$ for a problem domain, a generation perturbative hyper-heuristic *GPH* generates a new low-level perturbative heuristic *lph*, using the heuristics and/or components in C with conditional branching and iterative constructs, to produce a new perturbative heuristic for either i or the problem instances in I and similar problems.

The low-level perturbative heuristics that generation perturbative hyper-heuristics have been used to generate include local search operators [10, 63], algorithms [49] to solve problems and meta-heuristics [88, 163], as discussed below.

N. Pillay, R. Qu, *Hyper-Heuristics: Theory and Applications*,
Natural Computing Series, https://doi.org/10.1007/978-3-319-96514-7_5

5.2 Generating Local Search Operators

Grammar-based genetic programming has been used to evolve local search operators [10, 63]. Existing human-derived local search operators for the problem domain are decomposed into components. For example, the Boolean satisfiability problem involves determining an assignment of true and false values to variables that results in the well-formed formula evaluating to true. One of the existing perturbative heuristics, GSTAT [63], chooses a variable in the formula with the highest net gain, while another perturbative heuristic, GWSTAT, randomly selects a variable in a randomly broken clause. Examples of the components that these heuristics are decomposed into include net gain, randomly select a broken clause, and return the formula.

A grammar is used to define how the components can be recombined using branching constructs to produce new heuristics that are syntactically correct. For example for the Boolean satisfiability problem, examples of conditional-branching constructs used include IF-RAND-LT and IF-TABU [63]. IF-RAND-LT takes a floating-point number and two variables as arguments. If randomly generated floating- point number is less than the floating point argument, the first variable is returned; otherwise the second variable is returned. IF-TABU takes an integer value representing *age* as input and two variables. If the age of the first variable is less than the age argument, the second variable is returned; otherwise the first variable is returned.

In the study conducted in [63], a new genetic operator *composition* is introduced which combines two elements of the population, i.e. two heuristics, into one composite heuristic.

Sabar et al. [164] employed grammatical evolution with an adaptive memory mechanism to create perturbative heuristics for combinatorial optimization problems. Each heuristic is a combination of acceptance criteria, neighbourhood structures and neigbourhood structure combinations. The neighbourhood structures are problem dependent, e.g. swapping two examinations for the examination timetabling problem. The neighbourhood combinations combine the neighbourhood structures, e.g. union which applies two neighborhood structures consecutively. The adaptive memory aims to maintain diversity by maintaining a population of solutions, which are initially created using a construction heuristic and regularly updated when improvements on the population are found. The generative hyper-heuristic was applied to the examination timetabling and vehicle routing problems, and performed competitively with state-of-the-art techniques in solving these problems.

5.3 Creating Algorithms and Meta-Heuristics

Low-level perturbative heuristics evolved by generation perturbative hyper-heuristics also include algorithms. Genetic programming has been used to generate algorithms to solve the automatic clustering and travelling salesman problems

[49]. Each candidate algorithm comprises standard algorithm constructs, namely, a conditional-branching construct, an if-then-else statement, an iterative construction, namely, a while loop, and the logical AND operator. These constructs are combined with problem specific terminals. For example, the terminals for the travelling salesman problem add cities to the tour, e.g. the *best neighbour* heuristic adds a city which is the closest to the last city added, and the *near centre* heuristic adds the city nearest to the central point.

Generation perturbative hyper-heuristics have also been used to evolve meta-heuristics. Linear genetic programming has been employed for this purpose [88]. The candidate meta-heuristics comprise components of the meta-heuristics and low-level perturbative heuristics for the domain. Conditional-branching constructs are combined with components of the low-level perturbative heuristics for the problem domain, e.g. for the travelling salesman problem IF2-CHANGE will apply the 2-CHANGE operator if this results in a shorter tour. The meta-heuristic components also include an iterative construct, namely, REPEAT-UNTIL-IMPROVEMENT. A grammar is used to specify the correct syntax of the generated meta-heuristics. The evolved meta-heuristics were found to perform better than hill climbing and greedy hill climbing in solving an instance of the travelling salesman problem. In a similar study [163], Cartesian genetic programming is used to evolve a memetic or iterated local search algorithm to solve the travelling salesman problem. Each algorithm contains a while loop with a sequence of existing low-level perturbative heuristics for the problem domain; each sequence is automatically generated using Cartesian genetic programming.

5.4 Discussion

This chapter presents generation constructive hyper-heuristics, which create new low-level perturbative heuristics. The low-level heuristics have taken the form of new local search operators and new algorithms for solving specific combinatorial optimization problems, and new meta-heuristics. As with generated constructive heuristics, the new perturbative heuristics can be disposable [10] or reusable [49, 63]. Disposable heuristics are created for a particular problem instance, while reusable heuristics are induced using a training set of problem instances and the generated heuristics can be used to solve other problem instances. The generated perturbative heuristics are composed of existing low-level perturbative heuristics or components thereof combined with conditional-branching constructs and/or iterative constructs. Generation pertubative hyper-heuristics have not been researched as thoroughly as other hyper-heuristics, and the domains that they have been applied to include the travelling salesman problem, the Boolean satisfiability problem and the automatic clustering problem. Table 5.1 provides an overview of these applications.

Table 5.1 Generation perturbative hyper-heuristics

Problem domain	Local search operator	Algorithms	Meta-heuristics
Boolean satisfiability	Bader-El-Den and Poli [10] Fukunaga [63]	- -	- -
Examination timetabling	Sabar et al. [164]	-	-
Vehicle routing	Sabar et al. [164]	-	-
Travelling salesman	- -	Contreras-Bolton and Parada [49]	Keller et al. [88] Ryser-Welsch et al.[163]
Automatic clustering problem	- -	Contreras-Bolton and Parada [49]	-

Chapter 6
Theoretical Aspect—A Formal Definition

6.1 Introduction

Along with the continuous developments in hyper-heuristic (HH), various descriptive definitions for HH have emerged, leading to classifications of HH. Initially, hyper-heuristics have been defined as a search technique "to decide (select) at a higher abstraction level which low-level heuristics to apply" [51], "to combine simple heuristics" [162], or recently as a search method or learning mechanism for selecting or generating heuristics to solve computational search problems [30]. HH is thus categorized into four classifications, namely, selection perturbative / constructive, generation perturbative / constructive (see Chapters 3, 2, 5 and 4). Some attempts have also been made to generalize these classifications of HH, to allow both selection / generation and offline / online learning to interoperate within a repository [180]. It has also been proposed that the "domain barrier" in the HH definition should be moved so more knowledge can be easily incorporated in a more expressive HH for inexperienced practitioners [179].

This chapter presents a formal definition of HH based on the existing conceptual definitions in the literature [35]. Within the two-level framework of HH, two search spaces, namely the heuristic space and the solution space, are under consideration. Some fundamental issues are then discussed within this framework. In addition to the different encoding and search operations, various objective functions are defined in both spaces to evaluate searches on heuristics and direct solutions, respectively.

Within the two-level framework of HH, a selection constructive hyper-heuristic is then demonstrated to illustrate the inter-relationship between the two search spaces. A landscape analysis on the heuristic space reveals interesting characteristics for designing more effective hyper-heuristics. Future potential research developments are finally presented based on existing research advances addressing the theoretical aspects of HH.

N. Pillay, R. Qu, *Hyper-Heuristics: Theory and Applications*,
Natural Computing Series, https://doi.org/10.1007/978-3-319-96514-7_6

6.2 A Formal Definition of Hyper-Heuristics

A hyper-heuristic HH can be defined as a search algorithm for solving an optimization problem P, whose decision variables are heuristics, rather than direct solution variables in the optimization problem p under consideration. To solve P, HH explores at a higher level a heuristic space H of heuristic configurations h, which at a lower level generate direct solutions s in the solution space S for problem p. Two search spaces can thus be defined, namely a heuristic space H of P and a solution space S of p, each associated with an objective function, within the two-level framework [155].

Definition 6.1. Within a two-level framework, a hyper-heuristic HH explores heuristic configurations $h \in H$ in the heuristic space H at a high level. The performance of HH is measured using $F(h) \rightarrow R$. At the low level, an objective function $f(s) \rightarrow R$ evaluates the direct solutions $s \in S$ in the solution space S for the optimization problem p under consideration.

Solution s is obtained by using a corresponding heuristic configuration $h \in H$, i.e. $h \rightarrow s$. Let M be a mapping function M: $f(s) \rightarrow F(h)$. The objective of HH is to search in H for the optimal heuristic configuration $h*$, which generates the optimal solution(s) $s*$, so that $F(h*)$ is optimized:

$$F(h* \mid h* \rightarrow s*, h* \in H) \leftarrow f(s*,\, s* \in S) = \mathbf{min}\{f(s), s \in S\} \qquad (6.1)$$

The following terminologies are defined in the formal HH definition [155].

- Problem p: an optimization problem under consideration, whose direct solutions $s \in S$ are evaluated against objective function $f(s)$.
- Problem P: an optimization problem considered by HH, whose decision variables are heuristic configurations $h \in H$ evaluated against objective function $F(h)$.
- Solutions s: direct solutions for p.
- Heuristic configurations h: configurations upon low-level heuristics in L for P.
- Solution space S: consists of s for p, obtained by using h, i.e. $h \rightarrow s$.
- Heuristic space H: consists of h for P, explored by high-level heuristic algorithms *HLH* in *HH*.
- Low-level heuristics L: a given set of domain specific heuristics configured by *HLH* at the low level to compose h, i.e. L contains the set of domain values for the decision variables in h.
- High-level heuristics *HLH*: search algorithms or configuration methods at the high level upon L to search $h \in H$ for P.
- Objective function f: fitness evaluation for p, i.e. $f(s) \rightarrow R$ evaluates $s \in S$, s obtained using $h \in H$.
- Objective function F: fitness evaluation for P, i.e. $F(h) \rightarrow R$ evaluates $h \in H$ explored by *HLH*. The objective is to find the optimal $h*$, which obtains the optimal solution $s*$ for p, i.e. $h* \rightarrow s*$.
- Mapping function M: $F(h) \leftarrow f(s)$: each h maps an s, thus the performance of *HLH* upon h is measured based on the evaluation of its mapping s. Note that

F(h) may not be the same as *f(s)* although it is the case in most of the existing *HH* literature.

In the HH literature, different optimization problems *p* can be solved by plugging in a problem specific set *L* at the low level. The design of HH can thus be focused on the design of the high-level *HLH*. Solving different *p* thus can be transferred to solving a general optimization problem *P*; the latter can usually be encoded with lower-dimension representation and is easier to explore [155]. The generality of HH is also raised, as problem specific details and constraint handling are left with the direct solutions *s* obtained at the low level for *p*. Due to the above definition of the two search spaces at the two levels, the burdens of designing problem specific algorithms are also eased, focusing on the high-level configurations of heuristics. HH showed to be easy to implement, and has been successfully applied to a wide range of combinatorial optimization problems [30].

In [151], a formal definition of selection constructive HH based on graph colouring is presented. The above formal definition is extended to define both types of selection and generation HH with constructive and pertubative *L* as classified in [31]. Note that the *p* at hand may be either continuous or discrete. In most of the current HH literature only combinatorial optimization problems are investigated. The formal HH definition can also be extended to define continuous optimization problems, which represents a new line of interesting future research directions.

6.2.1 Two Search Spaces Within the Formal Hyper-Heuristic Framework

HH solves *p* by indirectly configuring and exploring *h* in *H* at the higher level, which then used to search for direct solutions *s* in *S*. Therefore, it is necessary to distinguish between the heuristic space *H* for *P* and the solution space *S* for *p*. In the current literature, most HH mainly explore $h \in H$, aiming to find h^* that maps to (near-)optimal $s^* \in S$, with less focus on the low-level *S*. Note, however, that within *S*, search can also be conducted by applying standard meta-heuristics to directly search *s* for *p* [151]. Table 6.1 compares the characteristics of the two search spaces based on terminologies defined for HH in Section 6.2.

Table 6.1 Characteristics of the two search spaces in the formal hyper-heuristic framework

Search Space	**Heuristic Space *H***	**Solution Space *S***
Encoding	Heuristic configurations *h*	Direct solutions *s*
Operation	High-level methods *HLH* upon the given *L* to configure *h*	Move or evolutionary operators on *s*
Objective Function	Evaluation function $F(h)$ upon *h* for *P*, $F(h) \leftarrow f(s)$	Objective function $f(s)$ on *s* for *p*

In selection hyper-heuristics, operations in HH often employ evolutionary algorithms or local search algorithms to configure h based on L [30]. Other configuration methods are also studied, including choice functions and case-based reasoning [30], see Chapters 2 and 3. In generation hyper-heuristics (see Chapters 4 and 5), genetic programming and its variants [11, 193] are often used to generate h, which can act as new problem specific heuristics to produce $s \in S$. Each h maps an s, thus the process of configuring or searching $h \in H$ simulates a search process exploring the mapping $s \in S$.

In most of the HH literature, $F(h) = f(s)$ [30]. Different evaluation functions, however, can be used in H and S, respectively. For example, in [17, 52, 119], a reward is used as F for a choice function to assess L and configure h at the high level, and a different problem specific evaluation function is used to evaluate the mapping s. Further in-depth studies may explore different M: $F(h) \leftarrow f(s)$ in the two-level framework, to design effective selection or generation HH with different high-level configuration methods and problem specific L.

HH indirectly searches $s \in S$ by exploring $h \in H$, thus may not directly explore from s towards the (local) optimal solutions $s^* \in S$, evaluated against f, as standard meta-heuristics usually do. Depending on the type of L in HH, neighbourhood solutions s' explored in S mapped by h' may or may not be the neighbourhood solutions from their precedent s mapped by h.

- In most HH employing perturbative L, individual low-level heuristics in h operate consecutively on complete direct solutions s, thus s can be seen as explored directly by the high-level search towards the (near)-optimal solutions s^* in S, guided by f on direct solutions s.
- In HH employing constructive L, solutions s and s' are constructed by h and h', thus s' obtained indirectly in S by h' may not be the neighbours of s, even if its corresponding h' is the neighbour of h. This is because during the solution construction using h', any different values assigned to variables in a partial solution using a different low-level heuristic in h', compared to h, are likely to lead to a different complete solution s'. Thus s' produced by the successive h' explored from h may not be neighbours of s in S.

Figure 6.1 presents the relationship between $h \in H$ and $s \in S$ within HH. In H, h_2 and h_3 are two successors of h_1 using an operation at the high level. In an HH that employs constructive L, their mapping corresponding solutions s_2 and s_3 in S may not be neighbours as defined using different (or even the same) operations upon the direct solution s_1 in S, obtained from h_1.

Given the characteristics in Table 6.1, it is noted that the size of H is very likely to be different from the size of S. In particular, S consists of all the possible direct solutions s for p, while H consists of heuristic configurations h for P. However, depending on how encoding and operation are defined, some of the s may not be obtained from any $h \in H$. This is reflected in Figure 6.1: s_4 may be a neighbour of $s_1 \in S$ using a specific operation; however, it may not have any corresponding $h \in H$ depending on how h is configured at the high level. In the example presented

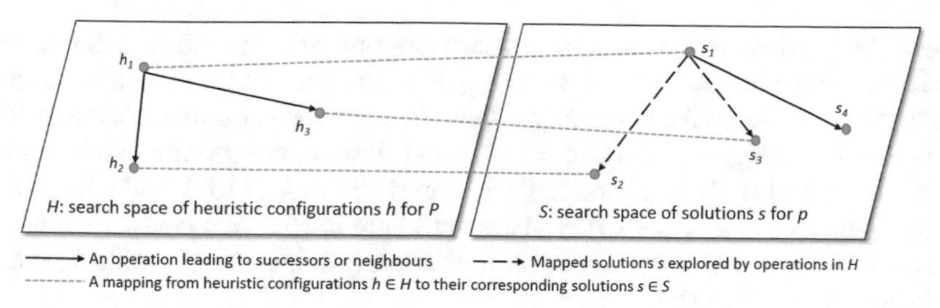

Fig. 6.1 Search in the two spaces H and S

in Section 6.3, this interesting issue has been investigated to explore search within both P and S to reach all $s \in S$ for p.

6.2.2 Fitness Landscape of the Heuristic Space in the Hyper-Heuristic Framework

In meta-heuristics, the concept of fitness landscape has been adapted to analyze the search space of solutions [115], revealing useful characteristics for designing more effective algorithms. For example, analysis of the landscape of the travelling salesman problem reveals an interesting feature called a "big valley", indicating a positive correlation between solutions and their fitnesses with the optimal s^*, i.e. solutions closer to s^* are of better quality [128]. This observation may be used to design effective encodings and operators to guide the search towards s^* in TSP and other problems with similar features in the landscape.

Based on the definitions of fitness landscape in state space theory [115], the fitness landscape of H for the optimization problem P in the HH framework (Section 6.2) can then be defined with three factors, namely an encoding using some finite alphabet in L to represent all possible heuristic configurations h; a successor operator to define how $h \in H$ are connected (explored), and a fitness function $F(h) \rightarrow R$ that assigns a fitness value R to each $h \in H$.

In some HH where h are encoded as one-dimensional sequences of low-level heuristics in L, it is possible and useful to conduct landscape analysis on H, whose spatial structure can be defined using the operation and a distance metric D on h. This proved to be very difficult, if not impossible, for $s \in S$ for many complex combinatorial optimization problems with n-dimensional solutions, n ≥ 2. In the literature, fitness distance correlation *fdc* has been mostly used to analyze landscape properties and measure problem difficulty. Given a set of encodings $h_1, h_2, ..., h_n$ and their fitness F, and the distance of $h \in H$ to their nearest optimum $h_{opt} \in H$, the *fdc* coefficient is defined as follows [87]:

$$fdc: \ \sigma(F, D_{opt}) = Cov(F, D_{opt}) / \sigma(F)\sigma(D_{opt}) \tag{6.2}$$

where $Cov(.,.)$ denotes the covariance of two random variables and $\sigma(.)$ the standard deviation. In the literature, the optimal h_{opt} is estimated by the h which leads to the best solution $s*$ found for p. For h_i, fdc thus indicates how closely their F and D are related to that of h_{opt}. A value of $\sigma = 1.0$ ($\sigma = -1.0$) for maximization (minimization) problems indicates F and D are perfectly correlated to h_{opt} [87], and thus provide perfect guidance to h_{opt}; thus P is an easy problem in HH. In a landscape analysis, a value of $fdc \leq 0.5$ ($fdc \geq 0.5$) for maximization (minimization) problems usually indicates an easy P in HH.

More landscape analysis could be conducted using other measures including auto-correlation [192], which calculates the fitness correlation of a series of h recorded along a random walk over a time series T. The longer the time lag between two correlated h in the random walk, the less rugged is the landscape of H thus the easier the problem for HH. This also indicates from another aspect the difficulty of search problem P in H.

6.3 Example: A Selection Constructive Hyper-Heuristic for Timetabling Problems

A graph-based selection constructive hyper-heuristic in [32] is re-defined in this section based on the formal definitions of HH in Section 6.2 for educational timetabling problems. Based on an analysis of the two search spaces, a hybrid HH [151] is demonstrated, together with a landscape analysis on H in this HH framework [127]. More details of the work can be found in the original papers [32, 127, 151, 155].

6.3.1 A Graph-Based Selection Hyper-Heuristic (GHH) Framework

In timetabling, graph colouring heuristics (see more details in Section 10.2) are constructive heuristics that order the events using some difficulty measure strategies. The ordered events are then assigned, one by one starting with the most difficult ones, to construct complete timetable solutions. The basic assumption is that the most difficult events need to be scheduled earlier to avoid problems at a later stage. For example, if *SD* (Saturation Degree) is used in an exam timetabling problem, the exams are ordered by the number of remaining valid slots in the partial timetable during the solution construction, and the most difficult one is scheduled first to avoid the problem of no valid slots left at a later stage.

A graph-based selection constructive hyper-heuristic (GHH) is defined as follows: On the high-level space H, a local search algorithm as the high-level heuristic HLH explores heuristic sequences $h \in H$ using the low-level graph colouring constructive heuristics in $L = \{LD, LWD, SD, LE, CD\}$, as explained in Section 10.2.

Each $h = \{h_1, ..., h_n\}$, $h_i \in L$, is evaluated by $F(h) \rightarrow R$. n is the problem size, i.e. the number of decision variables in p.

At the lower level of GHH, a timetable solution $s \in S$ is constructed iteratively by using an $h \in H$, considering constraints and f for the timetabling problem p (see Appendix B.4). At iteration i, $h_i \in h$ is employed to order the events not yet scheduled in p using its corresponding ordering strategy. The first event in the ordering (i.e. the most difficult one using h_i) is then scheduled in s. In the next iteration, h_{i+1} in h is used to reorder and schedule the most difficult remaining events. This process is repeated until a complete s is constructed. Any h that leads to infeasible solutions is discarded. The objective function $f(s) \rightarrow R$ evaluates $s \in S$ for p (see Appendix B.4).

The mapping function is defined as M: $F(h) = f(s)$, $h \rightarrow s$. The optimization problem P in the HH framework is thus to search for h^* of L at a higher level in H which constructs (near-)optimal solution(s) s^*.

6.3.2 Analysis of Two Search Spaces in the GHH Framework

In the GHH defined above, different local search algorithms are employed at the high level [151] to search for $h \in H$, and a greedy steepest descent method is used at the low level to exploit local optima from $s \in S$; s is obtained using the corresponding h. Thus search has been conducted within both H and S, with characteristics given in Table 6.2. Note that different meta-heuristics can be employed at both levels, and the objective functions at the two levels can be different.

Table 6.2 Characteristics of the two search spaces in the GHH framework

Search Space	**Heuristic Space *H***	**Solution Space *S***
Encoding	Sequences of heuristics h	Direct timetable solutions s
Upper Bound of the Search Space	n^e (e: length of h; n: size of L, i.e. $\|L\|$)	t^e (t: no. of slots; e: no. of events)
Operator	Randomly change two h_i in h	Move events in s to new slots
Objective Function	Cost of s constructed by the new h	Cost of the new neighbour s

Within GHH, the high-level search explores h rather than direct solutions s. As an s is constructed by an h step by step, similar neighbouring h in H may construct quite different s, likely to be widely distributed in S. As illustrated in Figure 6.1, by making local neighbourhood moves from h_1 to h_2 or h_1 to h_3 at the high level in H, GHH can explore s_2 or s_3 across very different regions in S. A local search at the low level upon s in S, on the other hand, usually generates similar local solutions, i.e. s_3 to s_4. The GHH search thus can be seen as exploring much larger neighbourhood regions in S using a local search in H, similarly to the search behaviour of large-neighbourhood search algorithms.

In [151], a fast steepest descent method is hybridized at the low level within GHH to further exploit a local optimum from $s \in S$. The motivation is twofold: First, the steepest descent in S can exploit local areas around s_3 to reach local optima quickly; Second, GHH thus is able to explore the whole search space S including s_4, which may not be reached by any $h \in H$.

6.3.3 Performance Evaluation of GHH

In [151], four different local search algorithms, namely steepest descent, tabu search, variable-neighbourhood search and iterated local search, have been used as the high-level search to explore h for both the course and exam timetabling problems, employing the same L, as presented in Section 6.3.1.

It was found that although variable-neighbourhood search and iterated local search performed slightly better, in general high-level search within GHH did not play a crucial role. This may be because that at the high level, the h are not concerned with the actual assignments of decision variables in s for p, but are indirect configurations of constructive heuristics, which are then used to build s. s sampled by h at the high level tend to *jump* within S; thus s and s' from the neighbouring h and h' are not successive neighbours. The different search methods used in H thus did not directly lead to different performance of HH upon S.

When employing steepest descent at the low level in S on each complete s constructed by h, GHH obtained significantly better results. Although the local optimum h in H at the high level might not map a local optimum s in S, the steepest descent upon s further explores S, leading to locally optimal solutions for p. Within GHH, the role of the high-level local search in H can thus be seen as to explore S indirectly, while the steepest descent search at the low level is to exploit local regions in S.

Penalties of timetable solutions obtained by GHH using iterated local search on the exam timetabling problems (see Appendix B.4) are presented in Table 6.3, compared against existing algorithms. In [151] exactly the same GHH is applied to both the exam and course timetabling problems. The only difference is $f(s) \rightarrow R$ on $s \in S$ for different p. Note that some of the existing approaches in Table 6.3 are not hyper-heuristics, and are specially designed for solving the specific problem under consideration, thus may not have been applied to solve both problems.

The overall idea of the exploration in H and exploitation in S using search at two levels in GHH is similar to that of memetic algorithms or genetic local search, where genetic operators applied to the population of solutions in S facilitate global exploration, while the local search on solutions in the population conducts exploitation within local regions. The difference is that GHH explores S by indirectly searching H, at a high level, in the manner of local search. The hybrid GHH is much simpler yet is capable of exploring and exploiting the search space S at two levels.

HH aims to increase the level of generality in solving multiple problems and problem instances, while most of the existing HH approaches have been applied to one problem domain, or evaluated by specific objective functions for different

Table 6.3 Penalties of timetable solutions by GHH on benchmark exam timetabling problems against existing approaches; details of the problem and penalty function can be found in Appendix B.4

	car91	car92	ear83 I	hec92 I	kfu93	lse91	sta83 I	tre92	ute92	uta93 I	yok83 I
GHH	5.3	4.77	38.39	12.01	15.09	12.72	159.2	8.74	30.32	3.42	40.24
LNS [2]	5.21	4.36	34.87	10.28	13.46	10.24	159.2	8.7	26	3.63	36.2
Fuzzy [6]	5.2	4.52	37.02	11.78	15.81	12.09	160.4	8.67	27.78	3.57	40.66
Adaptive [39]	4.6	4.0	37.05	11.54	13.9	10.82	168.7	8.35	25.83	3.2	36.8
Local search[25]	4.8	4.2	35.4	10.8	13.7	10.4	159.1	8.3	25.7	3.4	36.7
Hybrid [42]	6.6	6.0	29.3	9.2	13.8	9.6	158.2	9.4	24.4	3.5	36.2
Heuristics [43]	7.1	6.2	36.4	10.8	14.0	10.5	161.5	9.6	25.8	3.5	41.7
Tabu Search [67]	6.2	5.2	45.7	12.4	18.0	15.5	160.8	10.0	29.0	4.2	42.0
Hybrid [114]	5.1	4.3	35.1	10.6	13.5	10.5	157.3	8.4	25.1	3.5	37.4

problems, respectively. Thus the generality of HH approaches have not yet been assessed using a uniform or consistent measure. In recent research, a performance assessment for HH, at four different levels of generality, has been proposed [147]. Such study shows a welcome attempt to address the fundamental aspects of further research developments in HH.

6.3.4 Fitness Landscape Analysis on GHH

In the literature some landscape analysis has been conducted using measures such as *fdc* and auto-correlation, both indicating from different aspects the difficulty of search in H. An example analysis using *fdc* as explained in Section 6.2 to analyze H in GHH is presented in this section. More details can be found in [127].

Based on a variant of GHH using two low-level heuristics, *LWD* and *SD* as defined in Section 10.2, in [127] the landscape of H in GHH has been analyzed to gain insight into the global structure of H. As in the literature, the best known h obtained, h_{opt}, is used as an estimation of the optimal solution in the *fdc* analysis (Equation 6.2). $h \in H$ can thus be represented by binary strings, whose distance D is measured using Hamming distance.

In the *fdc* analysis, locally optimal h are measured against h_{opt} to reveal landscape features of H, indicated by the correlations between their distances and costs. A set of h is first randomly generated, one of each distance j, $j = 1, ..., l$, away from h_{opt}, l is the length of h_{opt}. A non-deterministic steepest descent search using one-flip neighbourhood moves is then applied 10 times to these binary h to generate 10 locally optimal h for each j. In total $LO = 10 \times l$ locally optimal h are thus obtained and their correlations with h_{opt} using (Eq. 6.2) are calculated. More details can be found in [127].

The fitness values of these LO for two example timetabling instances, *hec92 I* and *sta83 I*, are plotted in Figure 6.2, ordered increasingly by their costs. The plots show a number of local optima of the same cost especially for *sta83 I*, demonstrating several plateaus in H.

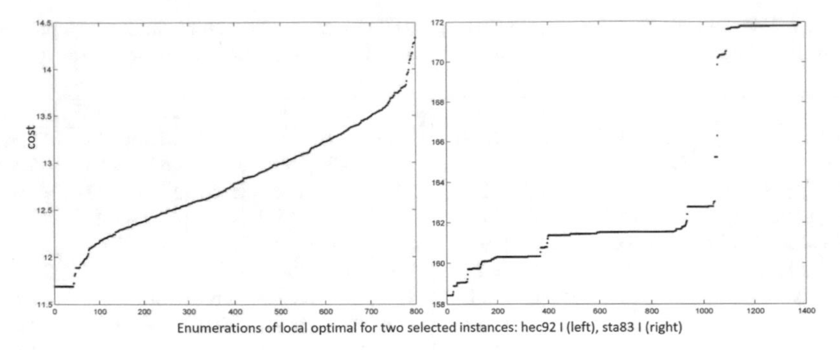

Fig. 6.2 Cost of local optimal h in GHH [127].

Figure 6.3 shows the best 10% of these *LO* local optima. Some interesting patterns can be extracted especially at the beginning of h. For example, values at certain positions h_i in the top h are fixed, i.e. the first four positions in h for *hec92 I* are always *LWD*. It is not surprising to see that random patterns appear at the end of h, as the last steps of solution construction tend to make less impact on the quality of s. No obvious patterns can be observed on lower-quality h.

Fig. 6.3 The best 10% local optimal h in GHH: hec92 I (left), sta83 I (right). Note: at each horizontal line white and black plots indicate low-level heuristics *LWD* and *SD*, respectively

The scatter plots between $F(h)$, $h \in LO$, and their distances to h_{opt} indicate a moderate-to-high positive correlation (in the range of 0.51 to 0.64). This is a very useful "big valley" feature in the landscape of H, similar to that of TSP observed in the literature [128], meaning better local optima are closer to h_{opt} in H. This also indicates that search in H is probably easier, as $F(h)$ of locally optimal h provides a useful indication of how close they are to h_{opt}.

Other patterns can also be observed, revealing some interesting features in the landscape of H. Although presenting similar positive correlation, the scatter plot for instance *sta83 I* shows several wide plateaus of the same cost f at different levels. In addition, those h with a cost below 38 (around $l/2$ bits away from h^*) are of low-quality, and show no clear correlation i.e. they are randomly located in H. More details can be found in [127].

Due to the simple one-dimensional structure of h, it is possible to conduct landscape analysis on H. This showed to be very difficult, if not impossible, for s in S for some problems investigated including the timetabling, vehicle routing and nurse rostering problems.

6.4 Discussion

Along with more recent advances in HH research addressing different theoretical aspects, more interesting research issues have emerged that require a formal definition of different types of HH in the literature. Based on the existing different conceptual definitions, a formal definition of HH [155] is presented in this chapter as an *optimization problem* to provide a unified fundamental basis for further explorations of emerging research directions in future HH research.

To demonstrate the formal HH framework, an existing selection constructive hyper-heuristic employing high-level local search algorithms [151] has been redefined, along with a landscape analysis for educational timetabling problems. Within the two-level HH framework, two search spaces, namely, the heuristic space H and solution space S can be explored separately, each with its own objective function. Landscape of the high-level search space H with one-dimension sequences of h showed the "big valley" feature by using the *fdc* analysis. The relation / mapping between the two search spaces is worth further investigation under the formal HH framework for the four types of HH [31] described in Chapters 2, 3, 4 and 5.

Based on this HH framework, several future research directions are worth further exploration.

HH aims to raise the generality of algorithms for solving multiple problems. This poses an interesting research question, namely how the No Free Lunch Theorem (NFL) applies to this new type of search algorithms. Some interesting discussions in [150] analyzed the conditions under which the NFL applies to HH. Based on the statement, if a set of fitness functions associated with problems p are *closed under permutation* [194], it would make no sense to find a solver for such p. However, such a set of problems usually represents a small fraction of the whole, thus there may be a free lunch developing HH approaches for not too large a set of problems. It would be interesting to carry out more in-depth analysis on NFL within the formal definition of the HH framework, to further explore the scope of p HH may address.

In [100], a runtime analysis is conducted on a selection HH using a randomized local search. It is shown that configuring a set of neighbourhood operators with an appropriate distribution is crucial, and is also problem dependent. It also shows that online reinforcement learning on configuring operators may perform more poorly than fixed distribution of operators in selection perturbative hyper-heuristics. A further investigation within the formal HH framework for both selection and generation HH with constructive and perturbative low-level heuristics would establish the theoretical foundations of HH as that of runtime analysis for evolutionary algorithms [100].

Performance measures on most of the current HH approaches in the literature are problem specific, even where multiple problems are concerned, when assessing the performance of HH. Some progress has been made in [147] to devise a new generality performance measure for HH of different levels of generality for different problems. With the aim of raising the generality of search algorithms, this new performance measure can be associated with the formal HH framework, to provide evaluations of different HH across various problems.

A unifying mathematical formulation for hyper-heuristics is proposed in [180], where by using a high-level controller, elements of heuristic design (both constructive and perturbative heuristic activities) compete for resources within a shared repository workspace to configure better heuristics. There is thus no distinction between online and offline activities, and heuristic activities interoperate based on information shared from other heuristics. The formal HH framework defined in this chapter could be integrated with this unifying framework, where the heuristic space H of P is explored by the high-level controller to solve multiple problems p.

Recent research on landscape analysis observes that multiple "sub-valleys" exist in a single big valley for TSP [128]. Similar analysis of the high-level H in the HH framework, where h is encoded as one-dimensional strings or sequences, might reveal more insights into the high-level landscape in HH, and inspire the design of more effective HH across different problems p.

Part II
Applications of Hyper-Heuristics

Chapter 7
Vehicle Routing Problems

7.1 Introduction

Vehicle routing problems (VRP) [72, 50, 186] represent one of the most investigated combinatorial optimization problems [134], due to the problem complexity and their potential impact on real-world applications especially in logistics and supply chains. The basic VRP involves constructing a set of closed routes from and to a depot, each serviced by a vehicle of certain capacity. In each route, a vehicle delivers the required demand in an ordered list of tasks for customers. The objective is to minimize the total distance, satisfying the capacity for all customers on each route. In some problem variants the number of vehicles used is also minimized. More details of the problem models and benchmark datasets are given in Appendix B.3.

In Operational Research, combinatorial optimization problems (see Appendix B) represent a subset of optimization, which consists of assigning discrete domain values to integer decision variables in a problem. As one of the mostly studied combinatorial optimization problems, the basic VRP is NP-hard [65, 101]. When modelling real-world VRP applications such as routing in transport logistics, the introduction of a wide range of constraints further increases the complexity. Variants of VRP thus have been studied in research integrating common features ranging from time windows, to capabilities to uncertainties of the tasks. Meta-heuristics integrating simple techniques as well as exact methods [185] thus represent one of the recent promising directions addressing both the complexity and realistic features in variants of VRP (see surveys for VRP [97, 186], VRP with Time Windows (VRPTW) [22], Capacitated VRP (CVRP) [68]) and Dynamic VRP (DVRP) [159, 136].

The most studied variants of VRP in HH (Appendix B.3) are reviewed in this chapter to obtain observations and draw conclusions about the performance of Hyper-heuristics (HH) for the problems. At the low level, a range of domain specific *llh*, which are either low-level heuristics in selection HH or components in generation HH, are adaptively selected (Section 7.2). At the high level, both *selection* and *generation* HH (Sections 7.3 and 7.4) have been developed to configure both

N. Pillay, R. Qu, *Hyper-Heuristics: Theory and Applications*,
Natural Computing Series, https://doi.org/10.1007/978-3-319-96514-7_7

types of these *llh*. Due to the generality of HH, some of the research developed HH approaches across different problem domains. Representative work along a line / series of research developments, with focus on VRP, has been reviewed. HH across multiple domains is discussed in Chapter 11.

In the rich literature of VRP research, a large number of VRP variants (VRP, VRPTW, CVRP, and DVRP, etc.) have been modelled, with benchmark datasets established when evaluating meta-heuristics. Although VRP has not been examined extensively in HH as in meta-heuristic algorithms, most of these widely studied benchmark VRP (see Appendix B.3) have also been considered in HH, and in some cases evaluated against meta-heuristics. Interesting observations that emerged are discussed in Section 7.5, leading to new lines of future potential research on real VRP applications with more real constraints or problem features.

7.2 Low-Level Heuristics for Vehicle Routing Problems

In the most commonly used solution encoding in the VRP literature, customers or tasks are usually represented as nodes in a graph representing the routing network. This graph-based representation is used in this chapter. Based on the classifications of HH [31], *llh* used in VRP are grouped into constructive in Section 7.2.1 and perturbative in Section 7.2.2. These *llh*, configured by the high-level heuristics in HH, operate upon nodes, which represent either customers or tasks. Most of these *llh* are usually problem specific, and have been extended or hybridized in HH for VRP variants with different constraints and features.

7.2.1 Constructive Low-Level Heuristics in Vehicle Routing Problems

Two types of constructive *llh* are used in the VRP literature for selection and generation HH, respectively.

- In selection HH, classic constructive heuristics such as the *Saving* heuristic by Clarke and Wright [48] have been employed to construct routes in VRP. In some HH approaches, perturbative heuristics are also used together with these constructive *llh* to further improve the generated solutions; see Section 7.3.
- In generation HH, problem attributes (state attributes or problem features) are used and combined using function operators or grammars to generate new heuristic functions (trees) or sequences of heuristic templates, see Section 7.4. These attributes and operators / grammars can also be seen as elements of constructive *llh*, which are configured by the high-level heuristics in HH to construct solutions for the VRP.

In selection constructive HH, the most commonly used constructive *llh* are summarized as follows. These heuristics and their extensions and hybridizations are widely used in the VRP literature.

- *Greedy*: insert randomly selected customers, subject to constraints, into those routes incurring the minimum cost
- *Saving* by Clarke and Wright [48]: merge two routes into one, based on the savings obtained on the distance of the resulting route
- *Insertion* by Mole and Jameson [123]: insert customers into a route that leads to the minimum cost resulting from the insertion
- *Sweep* by Gillett and Miller [69]: subject to the capacity of vehicles, form clusters of nodes by rotating a ray through (i.e. sweeping) customers clockwise or anticlockwise from the central depot; each route is then built by considering each cluster of nodes as a travelling salesman problem
- *Ordering*: heuristics that order and insert customers into the routes using certain criteria, i.e. increasing / decreasing demand, and farthest / nearest to the depot, etc.

In generation constructive HH, each heuristic is represented as a tree or grammar, with terminals (i.e. problem attributes) combined by function operators (i.e. internal nodes) or grammars. These constructive heuristics are generated either online or offline based on a set of VRP training instances [175, 104], and used to choose nodes in the network when constructing routes for the VRP. The function operators (see Table 7.1) or grammars are usually general across different problems; see also Chapter 9 for packing problems. Terminals (problem attributes) are usually problem dependent; we show below the ones widely used in HH for VRP.

- *demand*: (expected) (normalized) demand of the task for a customer
- *capacity / load*: (normalized) capacity when leaving the pickup or delivery node
- *cost*: cost of delivering the demand for a customer
- *distance*: distance from the current node or the depot; average distance of the current nodes to the remaining nodes; standard deviation of distance to the remaining nodes; and the total distance of routes, etc.
- *time* related: for a node, (normalized) start *stw* and end *etw* of the time window; arrival *at* and departure *dt* times; service time *st*; wait time *wt* at the pickup or delivery node; and time *t* from the current node or depot; etc.
- *satisfied*: proportion of nodes already being served
- *depotCost*: cost to reach the depot from the node
- *most_constrained*: most constrained tasks by demand or capacity
- *angle*: in relation to the *Sweep* heuristic, the angle between the current node and the depot
- *density*: node density in the routing network graph

Table 7.1 Most used function operators in genetic programming

Function	Operation
Addition (+)	addition, which adds values of two child nodes
Subtraction (−)	subtraction, which subtracts values of two child nodes
Multiplication (∗)	multiplication, which multiplies values of two child nodes
Protected division (/)	division, which divides values of two child nodes; when denominator is zero a protected operation is used
Relational operators ($\leq, <, >, \geq, =, \neq$)	comparison, which returns *true* or *false* based on the values of two child nodes
(*exp*)	exponential function, which returns e^x, where x is the value of the child node
(*max*) or (*min*)	returns the maximum or minimum value of two child nodes
(*angle*)	angle between the coordinates of the node and the origin of the polar system (usually the depot)

7.2.2 Perturbative Low-Level Heuristics in Vehicle Routing Problems

Widely used perturbative operators in meta-heuristics have been used as *llh* in selection HH, in some research hybridized with constructive *llh*, for variants of VRP. In the literature, there is no existing research employing perturbative *llh* in generation HH. In most selection perturbative HH approaches, a subset or extensions of the following perturbative *llh* have been employed. Operations of these *llh* upon the routing solutions are usually subject to the constraints in the VRP.

- *shift*: move a single node to a different route
- *swap*: swap two adjacent nodes in a route
- *interchange*: swap two nodes from different routes
- *or-opt*: move consecutive nodes to a different position in the same route
- λ*-opt*: exchange λ edges in a route
- *Van Breedam* [66]: relocate, exchange and cross strings between two routes
- *crossover*: exchange (part of) routes between two solutions
- *ruin and recreate* (*remove and re-insert*, or *destroy and repair*): remove a number of nodes using some criteria (time or location based), and insert them back into selected routes using heuristics. A new route is opened for nodes that cannot be re-inserted due to constraints.

7.3 Selection Hyper-Heuristics for Vehicle Routing Problems

In the existing selection HH, a given set of perturbative *llh* are usually selected to improve initial complete solutions. In some approaches, both constructive and perturbative *llh* are selected to construct and then improve the VRP solutions within one HH framework.

7.3.1 Selection Hyper-Heuristics Using Perturbative Low-Level Heuristics

A variety of techniques have been employed in HH to select perturbative *llh* for VRP. These range from local search [149, 191], to classifiers [8], to a multi-armed-bandit mechanism [166]. The selected perturbative *llh* (Section 7.2.2), using online or offline learning, or by solution evaluations, are applied to iteratively improve complete solutions for variants of VRP.

Although not named HH, an adaptive large-neighbourhood search at the master (high) level within a unified framework is developed in [149] to address a unified pickup and delivery problem with time windows for five variants of VRP. Using roulette wheel selection, simple *destroy and repair* heuristics (see Section 7.2.2) compete to modify a large number of variables in complete solutions based on the scores adjusted during online learning. The general framework has improved a large number of best results across all five different variants, demonstrating a highly effective and robust approach with little tuning effort for large-scale real-world VRP with mixed constraints and features.

VRP is one of the combinatorial optimization problems in the problem library provided by HyFlex [29], see Appendix A.1, with different types of perturbative *llh* (*mutation*, *ruin and recreate*, *local search* and *crossover*). An empirical study is conducted in the HyFlex framework [177] using an iterated local search selection HH with multi-armed bandit for VRPTW variants (Solomon, Gehring-Homberger, see Appendix B.3), as well as course timetabling problems. Statistical analysis and fitness-landscape-probing techniques are used on a set of training instances to identify a compact subset of the eight most effective *llh*. It is found that operators' evolvability (number of neighbours with better or equal fitness) can be used as an indicator to distinguish and select *llh* within HH. Using HyFlex, an iterated local search is employed in [191] as the high-level search to select from the 12 perturbative *llh* using online learning. In [8], offline apprenticeship learning is used in HyFlex to train classifiers based on small VRP instances to select from the 10 *llh* to improve solutions for unseen instances.

In [166], VRPTW problems are decomposed into sub-problems. Sub-solutions generated using column generation are combined and improved by a selection HH by selecting from seven perturbative *llh* for the benchmark Gehring-Homberger VRP (Appendix B.3. A multi-armed-bandit method is used based on online accumulative rewards, and solutions are accepted using a Monte Carlo mechanism.

7.3.2 Selection Hyper-Heuristics with Both Constructive and Perturbative Low-Level Heuristics

Some of the selection HH approaches employ meta-heuristics at the high level to select both constructive and perturbative *llh*. The high-level heuristics usually explore or configure sequences of *llh* operators, which are applied to construct and iteratively improve direct VRP solutions.

In [113], heuristic rules are used in a multi-agent system to select perturbative *llh* to improve solutions for distance-constrained VRP. Within an agent meta-heuristic framework, a coalition of agents explore the search space concurrently. In addition to learning by individual agents, the agents are also improved by exchanging information based on collective online learning. The *llh* operators are grouped as intensifier (improvement) or diversifier (generation, mutation and crossover) operators to strike a balance in the search. As the operators are used in both the initialization and during optimization, the HH can be seen as configuring and hybridizing both constructive and perturbative *llh*.

An evolutionary algorithm is developed in [66] at the high level to evolve sequences of constructive-perturbative *llh* pairs of variable lengths for 21 instances of Kilby's benchmark dynamic VRP [94]. Three types of *llh* are used, namely an ordering *llh* to rank customers, four constructive *llh* to construct solutions, and four repair heuristics to improve solutions. It is found that the *llh* evolved vary considerably for static and dynamic parts of the problem. It is crucial to design simple and effective *llh* considering adaption, average performance and speed. A diverse set of simple *llh* is recommended to improve the coordination among the *llh* during the evolution based on the communication of information on problem partial states.

In [122] a selection HH using an evolutionary algorithm is developed to evolve sequences of action units of *llh* (problem specific variations and mutations) like those in [66] for CVRP. It is observed that the best-evolved sequences of *llh* are composed of multiple actions of variations and mutations, indicating that larger neighbourhood structures are more effective at escaping local optima and producing high-quality solutions.

7.4 Generation Hyper-Heuristics for Vehicle Routing Problems

Compared to selection HH, generation HH aim to generate new heuristic functions or rules based on given *llh*. The newly generated functions or heuristics are then applied when solving new problems. The pre-defined set of *llh* usually includes problem attributes or features that are combined or configured using function operators or grammars by the high-level heuristics or methods. In the current literature, the most used high-level configuration methods for VRP are genetic programming [84, 175, 193, 104] and grammatical evolution [57, 164], using problem attributes (see Section 7.2.1) and genetic operators as shown in Table 7.1 for variants of benchmark VRP.

Genetic Programming HH In a two-level HH using GP, heuristic configurations at the high level are represented by trees, where terminal nodes representing problem attributes are connected by internal nodes representing function operators. The role of GP is to find the best heuristic configurations, usually by conducting training on a small subset of problem instances. In VRP, the design of GP consists of choosing a subset of the function set in Table 7.1, and the problem attributes as the terminal set (see Section 7.2.1). In [84, 193, 104] GP employs standard sub-tree crossover and mutation to evolve via generations new constructive heuristics based on selected training instances. The newly evolved heuristic functions or rules are applied to new instances of the same problem. Table 7.2 summarizes GP HH approaches for VRP.

Table 7.2 GP HH for VRP variants

	[193]	[104]	[84]	[175]
Terminals set	six attributes of vehicle behaviours	seven problem attributes	11 problem attributes	20 attributes and nine route selectors
Function set	$+,-,*,/,exp$ $max, sin, angle$	$+,-,*,/,min$ max	$+,-,*,/,max,exp$	$compare$
VRP variant	static and dynamic CARP	uncertain CARP	DVRP (Saint-Guillain in Appendix B.3)	CVRPTW (Solomon in Appendix B.3)
High-level h	Trees as heuristic functions h to calculate and assign the next vehicle (i.e. value for the next decision variable) until a complete solution s is constructed			
Low-level s	The direct solution, i.e. a set of routes, constructed using h			

In [193] the newly generated mathematical function is used to construct solutions of tours for five sets of capacitated arc routing problems (CARP), a counterpart of VRP where arcs rather than nodes are served by the vehicles. Six vehicle behaviours including demand, load, cost, satisfied, and depotCost, etc. (see Section 7.2.1) are used in the terminal set, configured and operated using eight mathematical operators (see Table 7.2, explained in Table 7.1) at the high level. The same problem objective function is used at both levels. The study also analyzed the features of the high-level search problem using two indirect representations in GP. The dimension of heuristic configurations is independent of problem size, and much smaller than that of direct solutions, as studied in [151]. The evolved new mathematical functions showed to perform well on both static and dynamic CARP.

A GP is developed based on training instances of Uncertain Capacitated Arc Routing Problems (UCVRP) with environment changes in [104]. The generated trees of new heuristics are used to order vehicle tasks during solution construction. Changes of task demand and edge accessibilities are addressed based on average and worst costs of the resulting solutions. Domain specific knowledge is studied to design effective GP HH. Three functions showed to be effective, by selecting a promising set of candidate tasks, detecting edge failures, and addressing route failures due to environment changes. Six function operators as shown in Table 7.2 are used, to configure problem attributes including demand, cost, load, depotCost, satisfied, and constant, etc.

Dynamic VRP with new arrival tasks is addressed in [84] using a GP HH. Average costs of training instances of nine different scenarios are used to measure the performance of HH to address the stochastic feature of real-time new task requests. In addition to a number of problem attributes including normalized travel time to the depot, normalized travel time from the current location, normalized service time, normalized demand, and vertex density, etc., two other terminals, the expected number of future requests and the probabilities of new requests, are also considered. The automatically generated heuristics are able to update routes with new arrival tasks, and significantly outperformed three manual heuristics. It is found that simple *llh* using probabilities as terminals does not improve the new heuristics generated.

In [175], GP is used to generate new constructive heuristics to construct solutions for a benchmark CVRPTW (Solomon in Appendix B.3) and a new real world VRP. A large number of problem attributes is considered in the terminal set. These can be catogerized into selecting nodes (average distance of the node to the remaining nodes, distance saving, first come fist served, slack in time window, time saving, and those in Section 7.2.1), and selecting routes (first route, least / most used route by time, route with most possible nodes, random route). Perturbative *llh*, local search and crossover operators are then selected by an iterated local search at the high level of HH to improve the initial solutions generated by the new heuristics. The newly generated heuristics alone showed to be competitive against seven standard constructive heuristics in VRP.

Grammatical Evolution HH In an HH using GE, high-level heuristic configurations are represented by a string genotype of variable length, i.e. a four-tuple Backus-Naur Form (BNF) <*T*, *N*, *S*, *P*> (see Table 7.3). Starting with a symbol in *S*, a production rule *P* in the form of a BNF grammar sentence is composed based on user-defined terminals *T* connected by operators or methods in a non-terminal set *N*. *P* can be recursively interpreted until all elements in *P* are terminals in *T*, to generate executable programs for the VRP. The generated programs (parsed trees) are evolved by using genetic operators in GE.

Due to the nature of indirect genotype encoding in GE, the BNF grammar can be easily defined by users and the resulting trees are evolved using high-level configuration methods for solving different problems. The newly generated grammar thus potentially could be more easily reused for other problem instances compared to the newly generated constructive heuristics by genetic programming. In the literature, however, GE is less studied for VRP compared to genetic programming in generation HH; see Table 7.3.

Table 7.3 BNF grammars in GE HH for VRP

BNF	**Sabar et al. [164]**	**Drake, Killis and Özcan [57]**
Terminals set *T*	Neighbourhood operators and 11 symbols (*AllMoves* and *Great Deluge*, etc.)	26 attributes in Section 7.2.1
Non-terminal set *N*	Acceptance criteria, LST configurations, Neighbourhood structures / combinations	12 arithmetic and relational operators in Table 7.1
Start symbol *S*	<*LST*>: Local Search Template	<*Initialization*><*Ruin*><*Recreate*>
Production *P*	Rules composed of *S* followed by elements in *N* and *T*	Rules / sentences consists of elements in *N* and *T*
Problem	DVRP (Christofides and Golden [72], exam timetabling (Carter, ITC2007 in Appendix B.4)	CVRP, VRPTW (Augerat, Solomon in Appendix B.3)

In [57], heuristics generated by GE are used not only to build the initial solutions for a variable neighborhood search (VNS) algorithm, but also to select *ruin and insertion* operators within the VNS for two VRP variants. Instead of systematically switching operators in a predefined order in VNS, GE is used to automatically select operators during the search. Another GE HH is devised in [164], using perturbative *llh* to automatically generate and evolve templates of neighbourhood operators and eight acceptance criteria for both dynamic VRP and exam timetabling problems. The GE is associated with an adaptive memory of high-quality solutions to increase diversity, leading to high performance in both problem domains.

7.5 Discussion

Current research in HH for VRP has addressed a diverse range of interesting issues. Both perturbative and constructive *llh* have been hybridised and selected in one selection HH, providing general methods applicable to real-world VRP with various features. In selection HH, studies on identifying a compact subset of effective *llh* revealed ideas to build more efficient HH frameworks. Combined *llh* (larger neighbourhood operators) in selection perturbative HH showed to be more effective at escaping from local optima. Online and offline learning for dynamic VRP obtained more insights on dealing with uncertainties in VRP, and increased the robustness and generality of HH. HH approaches serve as the mechanisms to explore and configure integrated approaches based on either online or offline learning, indicating promising future developments across both HH and meta-heuristics.

On the other hand, although generation constructive HH has been successfully applied to generate new heuristics or functions for variants of VRP, more research needs to be conducted to obtain in-depth analysis on generating new effective constructive and perturbative heuristics for different instances or problem variants.

Compared to other less-studied problem domains such as nurse rostering, VRP variants have attracted relatively more attention in HH partially due to the wide range of different constraints and problem features. Due to the high demand for its real-world applications, however, there is much more potential in HH to provide general solution methods for variants of VRP.

In VRP, perturbative *llh* configured by different selection HH approaches (evolutionary algorithms, iterated local search, and classifiers, etc.) have showed to be applicable directly to different VRP instances or variants. Automatically generated new constructive heuristics configured using offline learning on training instances based on problem state features, however, raise the question of reusability to variants of VRP. More analysis is needed on how the same generation HH framework with different constructive *llh* could be reused, with offline or online learning, to provide more insight to address this issue.

More findings of *llh* can facilitate more general and effective HH for a wider range of VRP applications. This chapter focuses on benchmark VRP with several different constraints or features in VRP variants. The high-level search in HH configures *llh*, which encapsulate details in the specific problem. A range of *llh* for VRP have been investigated, either on problem attributes with function operators in genetic programming or grammatical evolution, general move operators in local search, or genetic operators in evolutionary algorithms. Based on emerged findings, some of these *llh* can be extended to encapsulate other problem specific features, constraints and uncertainties, and applied to real VRP. A library of simple elementary *llh*, categorized to address specific aspects of vehicle / customer behaviour and constraints, focus of search (intensification or diversification), and acceptance criteria, and easily portable and extendable (i.e. via the form of XML) can facilitate designing such more effective and general HH frameworks for real-world VRP.

Chapter 8
Nurse Rostering Problems

8.1 Introduction

Personnel scheduling problems arise from various real-world scenarios, including supermarket staff scheduling, call centre staff allocation, police force scheduling, and, the most studied, nurse rostering in hospitals. Due to the demands of quality healthcare, limited resources, and the tight constraints of specific legislation worldwide, the nurse rostering problem (NRP) has received extensive research attention in the last five decades [26].

The NRP consists of assigning a set of nurses with different skills to a set of shifts of different types on each day or timeslot of a scheduling period, satisfying a set of constraints including coverage, legislation, personal preferences, and problem specific requirements. The objective is to minimize the violations of soft constraints (which ideally should be avoided, i.e. nurse preferences) while satisfying hard constraints (which must be satisfied, i.e. all shift demands must be covered during the scheduling period).

As an NP-complete problem [130], nurse rostering presents a research challenge in operational research and meta-heuristics. Algorithms or methods investigated include mathematical programming in operational research, evolutionary algorithms in artificial intelligence and hybrid approaches across different disciplines. Current research in HH on NRP has led to some interesting results based on the extensive research on meta-heuristic algorithms [30].

Along with the extensive NRP research in the last five decades, benchmark NRP datasets have been established, representing a good coverage of real-world problems with different features. These NRP datasets have been applied in HH research; more details can be found in Appendix B.2.

- *UK* dataset: One of the early datasets is derived from a major UK hospital, where a set of 411 pre-processed valid shift patterns / sequences is defined with associated costs calculated based on various constraint violations.
- *INRC2010* datasets: The First International Nurse Rostering Competition [76] (INRC2010) also established a set of benchmark datasets, aiming to bridge the

N. Pillay, R. Qu, *Hyper-Heuristics: Theory and Applications*,
Natural Computing Series, https://doi.org/10.1007/978-3-319-96514-7_8

gaps between theory and practice and promote advances of a range of new approaches.
- *Nottingham* datasets: An NRP web site has been established at the University of Nottingham, providing a collection of a range of NRP problems derived from hospitals worldwide, and the lower-bound solutions reported in the literature.

The majority of current HH approaches for NRP are selection based (Section 8.3), using a diverse set of techniques at the high level to configure a large set of low-level heuristic (*llh*) perturbative operators associated with different acceptance criteria for several well-established NRP benchmarks widely used in the meta-heuristics community. Based on the success of some meta-heuristic algorithms implemented for real-world NRP, more advances in both meta-heuristics and HH might further address the gaps between research and practice for this highly constrained combinatorial optimization problem.

8.2 Low-Level Heuristics for Nurse Rostering Problems

Due to the hard constraints in NRP, perturbative *llh* operators are usually defined subject to the fixed coverage requirement of specific shifts on the same day. Swaps of shifts, either consecutive or not, are made between selected nurses on the same day subject to the nurses' skill types. Different methods employed a different subset of the following *llh* and their extensions in the literature [119, 8, 76]:

- *change shift*: change the shift type of a (randomly) selected nurse based on his / her skill type.
- *swap shifts*: between two nurses, swap shifts on *n* consecutive or non-consecutive weekdays or weekends. Nurses may be randomly selected, or heuristically selected based on the number of their conflicts with others, subject to their shift types or skills.
- *move shift*: move the shift of a nurse to another nurse using certain criteria (randomly without considering the costs, or heuristically with the least cost incurred).
- *ruin and recreate*: un-assign and re-assign all shifts of a set of selected nurses in the roster solution randomly or heuristically.

The above list is not exclusive, but presents the most widely used *llh* in HH for NRP. These *llh* have been used with different settings, mostly as simple operators, to gain useful insights into their impact on the performance of different HH approaches. Note that some of the *llh* of smaller size (which make smaller changes to problem solutions) may be redundant given the other larger *llh* (i.e. one larger *llh* operation may be equivalent to the application of several smaller *llh* operations), but both types have been found to contribute to the flexibility of search at different stages of problem solving for instances of different landscapes.

Along with perturbative operators, acceptance criteria have also been investigated and compared in experimental studies [119]. Improving moves are usually accepted, while worsening moves are accepted using various criteria, to strike a balance between exploration and exploitation. Most widely used move acceptance criteria in meta-heuristics are not problem specific, thus can be easily employed in HH.

- *All Move* or *Naive Acceptance*: All neighbourhood solutions by each *llh* are accepted.
- *Only Improving*: Worse neighbourhood solutions are not accepted to encourage exploitation. This can be *first improvement*, which accepts the first better neighbour obtained, or *best improvement* to accept the best among a set of neighbours.
- *Improving or Equal*: Neighbourhood solutions of equal and better quality are accepted.
- *Late Acceptance*: A solution better than the last *n* previously visited solutions is accepted.
- *Simulated Annealing*: Worse solutions are accepted with a probability dependent on the difference from the neighbourhood solutions, and a temperature parameter. The probability gradually decreases, accepting less worse solutions at later stages of the search.
- *Great Deluge*: Worse solutions within a threshold *t* are accepted, and *t* is reduced during the search. Various strategies can be used to gradually reduce the threshold.
- *Adaptive Iteration Limited Threshold Accepting*: This criterion checks a list of recent neighbourhood solutions, and accepts new best solutions found after a number of worse moves, or uses fitness values of recent moves as a threshold for accepting worse solutions.

In HH, a subset of the above perturbative operators, associated with different acceptance criteria, have been used as operator-acceptance *llh* pairs, and selected in selection HH adaptively [17, 119, 8]. In other selection HH approaches, a fixed acceptance criterion is used at the high level, and only perturbative operators are considered and selected as *llh*. In some selection perturbative HH, these simple operators are integrated with acceptance criteria as one combined *llh*, and selected by the high-level heuristics [36, 52, 12, 4].

8.3 Selection Hyper-Heuristics for Nurse Rostering Problems

A diverse set of high-level methods have been used in selection HH, mainly on three benchmark datasets detailed in Appendix B.2, namely the UK dataset with 411 shift patterns, the Nottingham benchmark dataset, and the INRC2010 competition dataset. These methods include choice functions, adaptive strategies, Bayesian network and local search algorithms, and provide interesting findings on selecting perturbative *llh*. A summary of different selection HH research in NRP is presented in Table 8.1; details explained in this section.

Early research in HH developed various approaches for 52 instances derived from a major UK hospital by selecting different perturbative *llh* as listed in Section 8.2.

Table 8.1 Selection perturbative hyper-heuristics for nurse rostering problems

	High-level method	***llh***	**Dataset**
Cowling et al. [52]	Choice Function	9 perturbative operators	UK
Burke et al. [36]	Tabu Search	9 perturbative operators	UK
Aickelin et al. [4]	Bayesian Network	rules to select the 411 shift patterns	UK UK
Bai et al. [12]	Simulated Annealing	9 perturbative operators	UK
Bilgin et al. [17]	Random, Choice Function, Dynamic Strategy	12 *swap shifts* and *move shift* operators, with 4 acceptance criteria	INRC2010
Misir et al. [119]	Two adaptive strategies	29 *swap shifts* and *move shift* operators, with 7 acceptance criteria	INRC2010 instances
Shahriar et al. [173]	Iterated Local Search based on tensor analysis	4 types of perturbative operators	Nottingham

A set of 411 pre-defined valid shift patterns has been obtained for this problem, and the *llh* selected by the high-level methods operate upon these shift patterns. The problem solutions thus are improved using *llh* to perturb the available patterns / sequences of shifts rather than individual shifts.

In [52], a choice function learns to select from nine perturbative *llh* with the *first improving* acceptance criterion as defined in Section 8.2 to operate upon the 411 shift patterns in the UK benchmark NRP . The evaluation at the high level therefore rewards each *llh* based on its online performance, i.e. the cost of the resulting roster solutions. This same set of *llh* has also been employed in [36], where *llh* are selected by a high-level tabu search within a unified HH framework for the same UK NRP and also benchmark course timetabling problems. In [12], a simulated annealing HH (SAHH) is hybridized with a genetic algorithm to exploit local optima more efficiently. Based on the performance of the acceptance ratio of the nine *llh*, the 411 shift patterns are selected. Results demonstrate the high efficiency of the SAHH approach compared to those in the literature.

An interesting approach is investigated in [4], where a Bayesian network is used as the high-level method to select a string of rules. Each rule chooses from the collection of 411 shift patterns to build roster solutions. Based on a set of training instances, an estimation of distribution algorithm is used to learn the probabilities of the rules that contribute to constructing high-quality roster solutions. The evaluation at the high level is thus a measure of the likelihood the rules lead to good quality rosters. This novel approach can be seen as a selection constructive HH that selects *llh* rules using a statistical model to combine shift patterns to construct high-quality roster solutions.

For the Nottingham benchmark NRP dataset with a large number of diverse problems at hospitals worldwide, a tensor-based machine learning technique is used in [173] to extract patterns from the performance of *llh*. An iterative multi-stage algorithm is then used to automatically select four types of *llh* operators (mutation, crossover, local search, ruin and recreate) with *improving and equal* and *naive ac-*

ceptance in the HyFlex framework [29] (see Appendix A.1) based on the knowledge learned by the machine learning model.

The INRC2010 competition attracted a new line of HH research, where a large number of perturbative *llh* are systematically investigated. In [17], three high-level selection methods, namely random selection, choice function, and dynamic heuristic set strategy, are used to select in total 12 variants of *swap shifts* and *move shift* operators described in Section 8.2. These *llh* are categorized in three groups, by days, weekdays and weekends, and *n* (non)-contiguous days, for two randomly chosen nurses. Results have been compared against solutions obtained from integer linear programming within similar computational time.

Another intensive study in [119] employs a *monitor* to manage two high-level selection methods, namely hill climbing and tournament selection, and seven acceptance criteria to select nine heuristic sets. In total 36 variants of the HH approaches with different settings for the high-level search and *llh* are studied to evaluate its diversification and intensification on 10 instances in INRC2010, and also another two healthcare problems. At the high level, the monitoring mechanism manages at the low level four types of acceptance criteria and in total 29 *llh* perturbative operators of different sizes (number of changes to solutions) and speed (execution speed when applied). The effects of heuristic selection and acceptance are analyzed based on the frequencies of *llh* being called and the number of new best solutions found by each *llh*.

The in-depth analysis in [119, 76] revealed some interesting research findings within selection perturbative HH for NRP. A large number of different perturbative operators and acceptance criteria have been automatically selected and combined to develop general algorithms across different problems. HH can be seen as serving as a general framework, where different elements at the two levels can be configured. It was found that it is crucial to combine *llh* of distinct characteristics that can work together. Such a simple framework can also offer a flexible analytical tool to support effective algorithm design in meta-heuristics.

8.4 Discussion

Based on the existing rich research in nurse rostering problems (NRP) using a variety of techniques and algorithms, some interesting findings have been obtained on selection perturbative HH approaches. The analysis conducted on simple *llh* of different behaviours, number of changes (sizes) and execution speed has led to a deeper understanding of their performance on different instances [76]. With developments along different lines of research on problem specific *llh*, classifications of NRP, and diverse sets of benchmark problems, HH can be further extended to address a wider range of research issues in the meta-heuristic communities for these highly constrained combinatorial optimization problems.

In selection perturbative HH, investigations have been conducted to examine the effect of *llh* elements including perturbative / move operators and the associated acceptance criteria, on benchmark NRP [76, 119]. Analysis of the synergies between different operators and acceptance criteria at the low level has provided insights into effective search. In this sense, HH can be seen as an analytical tool and framework, to support in-depth analysis of different elements within local search algorithms. Research findings on the combinations among perturbative *llh*, in conjunction with different acceptance criteria, can be extracted to facilitate advanced design of efficient local search algorithms.

As highly constrained combinatorial optimization problems, various NRP have been investigated with a wide range of constraints and problem specific features from different countries. Compared to the diverse research using a variety of high-level algorithms for vehicle routing and examination timetabling problems (Chapters 7 and 10), current research in HH for NRP has mainly focused on selection perturbative approaches. There is a lack of research on constructive HH for NRP. More developments of constructive HH may require further research findings on effective modelling of the complex NRP problems, hybridized with perturbative approaches. For example, for the UK NRP dataset (Appendix B.2.2), a set of valid shift patterns are pre-defined considering various constraints. Selection HH can thus be effectively applied. For the highly constrained NRP, developing generation HH to generate new effective heuristics presents more challenges compared to selection HH; thus simple heuristics and operators are easier to apply to different instances and problems.

Based on the existing rich literature in NRP, HH can also be advanced by further extending *llh* addressing different constraints in different NRP. Research effort has already been made to categorize *llh*. Classifications of NRP have also been proposed [44], using similar notations to those used in the scheduling literature. Given the highly constrained nature of NRP, problem specific simple elementary *llh* could be classified according to which and how many constraints thy address. Both the HH and meta-heuristics communities may benefit from such a systematic study of the synergies between different categories of constructive and perturbative *llh*.

The extensive study of meta-heuristics for NRP has motivated the establishment of several diverse benchmark datasets; the findings obtained have also motivated HH research. The pre-processing of the dataset from a UK major hospital has demonstrated the effectiveness of constraint handling, by encapsulating the problem complexity into a collection of pre-defined valid shift patterns. Similar studies in the NRP literature [83, 188, 23] have also showed promising results using pre-defined valid shift sequences (also called workstretches or stints) of high quality, i.e. with fewer or no violations of constraints. The other two benchmark datasets, the NRP benchmark site maintained at the University of Nottingham (see more details at http://www.schedulingbenchmarks.org/), and the three tracks of datasets at the first nurse rostering competition INRC2010 provide lower bounds as well as a unified format of the problem description. Such efforts are highly valuable and are strongly encouraged to promote future advances in both HH and meta-heuristics communities.

Chapter 9
Packing Problems

9.1 Introduction

The packing of items into a container or bin so as to minimize cost is a common problem experienced in industry. This chapter examines the use of hyper-heuristics for solving bin packing problems presented in Appendix B.1. Selection constructive hyper-heuristics and generation constructive hyper-heuristics have been successfully employed to solve bin packing problems. The use of selection constructive hyper-heuristics is presented in Section 9.2 and generation constructive hyper-heuristics in Section 9.3. The chapter concludes with a discussion on hyper-heuristics for packing problems, including future research directions.

9.2 Selection Constructive Hyper-Heuristics

Selection constructive hyper-heuristics have been effective in creating initial solutions for bin packing problems. The aim of selection constructive hyper-heuristics, like low-level constructive heuristics, is to produce a good initial solution. This section provides an overview of using selection constructive hyper-heuristics for the one-dimensional bin-packing problem. Selection constructive hyper-heuristics select constructive heuristics to apply at each point in constructing a solution for the problem. In the context of bin packing problems this involves selecting a heuristic to choose a bin to place the next item in [105, 140, 161, 162]. Some of these heuristics by definition also choose the item to place in a bin, e.g. first-fit decreasing. However, the hyper-heuristic can select a heuristic to choose both the item and the bin in constructing a packing solution [140].

In applying a selection constructive hyper-heuristic, one has to decide on:

- The low-level constructive heuristics used for constructing solutions s.

N. Pillay, R. Qu, *Hyper-Heuristics: Theory and Applications*,
Natural Computing Series, https://doi.org/10.1007/978-3-319-96514-7_9

- The high-level method to select heuristic configurations, where each low-level constructive heuristic is used at each point in constructing a solution s for problem P.

The following section describes the low-level constructive heuristics used in selection constructive hyper-heuristics for bin packing problems. Section 9.2.2 describes methods that have been employed by selection constructive hyper-heuristics in solving bin packing problems.

9.2.1 Low-Level Constructive Heuristics for Bin Packing

There are essentially two types of low-level constructive heuristics for packing problems, namely, heuristics that choose the bin to place the next item in, and heuristics that choose the next item to be placed in the bin. Some heuristics combine both of these functions. For example, the first-fit decreasing heuristic chooses the largest item and places it in the first bin that it fits into. The low-level bin selection constructive heuristics used in selection constructive hyper-heuristics [105, 140, 161, 162] for bin packing problems include:

- First-fit (FF) - This heuristic places the item in the first bin that it fits into.
- Best-fit (BF) - The item is placed in the bin with the least amount of remaining space once the item is placed.
- Next-fit (NF) - The item is placed into a new bin if it does not fit into the current bin.
- Worst-fit (WF) - The item is placed into a bin with the most remaining space.
- First-fit decreasing (FFD) - The same as first-fit but items are sorted in descending order according to their size, and are allocated in order.
- Best-fit decreasing (BFD) - The same as best-fit but items are sorted in descending order according to their size, and are allocated in order.
- Next-fit decreasing (NFD) - The same as next-fit but items are sorted in descending order according to their size, and are allocated in order.
- Filler - This heuristic sorts the items to be placed in decreasing order according to their size, and attempts to allocate as many items as will fit into the existing bins. If none of the items fits, as many items as possible are placed in a new bin.
- Djang and Finch (DJD) - A bin is filled with items until it is one-third full, with the largest items chosen first. Combinations of one to three items are tried to fill the bin to its full capacity. If this is not successful, different combinations are tried to fill the bin to within 1% of its capacity and then 2% of its capacity, and so on. Variations of this heuristic have been used, such as Djang and Finch with more tuples (DJT), which is the same as the DJD heuristic but combinations of items from one to five instead of one to three are considered when attempting to fill a bin.

The first-fit decreasing, best-fit decreasing and next-fit decreasing heuristics decide on both the item to place next and which bin to place the item in. The following item selection heuristics have been defined specifically for selecting an item [140]:

- Largest item - The largest item is chosen to be placed to a bin.
- Availability degree - The first item that fits into an existing bin is chosen.
- Saturation degree - The item that has the smallest number of bins available that it can fit in is chosen next.

A selection constructive hyper-heuristic selects a heuristic to choose a bin, or a heuristic to choose an item and another heuristic to choose a bin, at each point in constructing a solution to the problem. The following section examines the techniques that have been used by selection constructive hyper-heuristics for bin packing problems.

9.2.2 Methods Employed by the Hyper-Heuristics

This section provides an overview of selection constructive hyper-heuristics employed to solve bin packing problems. Research in this area has basically taken one of two approaches, namely, generating condition-action rules [105, 161, 162] or heuristic combinations [140].

In the study conducted by Ross et al. [161, 162] the condition-action rules are generated to select a constructive heuristic to select a bin to place the next item in. The condition component of the rule defines the state of the problem in terms of:

- The number of huge items that still need to be packed. Huge items are those that have size larger than half the bin capacity.
- The number of large items that still need to be packed. Large items are those that have size in the range of a third to half the bin capacity.
- The number of medium items that still need to be packed. Medium items are those with size in the range of a quarter to a third of the bin capacity.
- The number of small items that still need to be packed. Small items are those with size less than a quarter of the bin capacity.
- The proportion of items that still need to be packed.

The action is the heuristic to be applied to choose a bin given the specified problem state. Each rule is composed of six numbers. The first five numbers are real values in the range 0 to 1 representing the problem state. The last number is an integer value corresponding to one of the constructive low-level heuristics. Ross et al. have used learning classifier systems (LCS) [162] and genetic algorithms (GA) [161]. The problem instances were divided into a training set and a test set. The training set was used by the LCS and GA to induce condition-action rules, and the best-performing rule was applied to the test set. In both studies the selection constructive hyper-heuristics performed much better than each of the low-level heuristics applied separately to solve the one- dimensional bin packing problem.

López-Camacho et al. [105] take a similar approach and use a genetic algorithm to evolve condition-action rules for selecting low-level bin selection constructive heuristics. As in the studies by Ross et al. [161, 162], the condition represents the state of the problem and the action the heuristic to apply. The problem state is defined in the same way as in [161] and [162]. The selection constructive hyper-heuristic was applied to the one-dimensional bin packing problem, the two-dimensional regular bin packing problem and the two-dimensional irregular bin packing problem. The problem instances for all three problems were divided into a training set and a test set. The best condition-action rule evolved by the genetic algorithm was applied to the test set of problem instances. The hyper-heuristic performed better than the individual constructive heuristics applied separately for all three problems.

In [140] the selection constructive hyper-heuristics use an evolutionary algorithm to explore the space of constructive heuristic combinations. The heuristics in each combination are applied in sequence to schedule an item, with one heuristic applied for each item. Two hyper-heuristics were tested. The first evolves combinations of heuristics to select a bin to place an item in. In the second hyper-heuristic each chromosome comprises two combinations. The first is a combination of heuristics to choose a bin and the second a combination of heuristics to choose an item. In this case the hyper-heuristic evolves the combination of bin selection heuristics and the combination of item selection heuristics simultaneously. Both hyper-heuristics produced better results than the low-level constructive heuristics applied separately for one-dimensional bin packing. The hyper-heuristic evolving both the bin selection heuristic combination and the item selection heuristic combination performed better than the hyper-heuristic generating just the bin selection heuristic combination. The combinations were evolved online for a particular problem instance.

9.3 Generation Constructive Hyper-Heuristics

Generation constructive hyper-heuristics have been found to be effective for bin packing problems. In implementing a generation constructive hyper-heuristic the researcher has to decide on:

- The operators that the new heuristics will consist of.
- The problem characteristics, existing heuristics and/or components of existing heuristics that will be included in the new heuristics.
- The method that will be used to combine the operators and problem characteristics and existing heuristics and/or components of the existing heuristics, into new constructive heuristics.

For the domain of bin packing the operators used are basically arithmetic and conditional-branching operators [82, 174]. These include:

- Addition (+): Performs the standard addition operation, which adds two values.

- Subtraction (-): Performs the standard subtraction operation, which subtracts two values.
- Multiplication (*): Performs the standard multiplication operation, which multiplies two values.
- Protected division (%): Performs the standard protected-division operation, which divides two values if the denominator passed to the operator is not zero. If the denominator is zero the operation evaluates to an integer value such as 1 [82] or -1 [174]. Due to the stochastic nature of the methods usually used to combine operators and problem characteristics, such as genetic programming [96], the denominator might evaluate to zero; hence a protected operator is used.
- Relational operators ($\leq$, $<$, $>$): Perform the relational operations between two values. For example, $\leq$ performs the less than and equal to operation. If the first argument is less than or equal to the second argument the operation returns 1, and -1 otherwise. Similarly, $<$ and $>$ return 1 if the first argument is less than or greater than the second argument respectively, and -1 otherwise.
- Conditional branching operators: These operators perform a function similar to if-then-else statements in a programming language. For example, the IGTZ [174] executes its third argument otherwise.

The problem characteristics used depend on the bin packing problem being solved. For example, problem characteristics used for the one dimensional bin packing problem include [82, 174]:

- The fullness of the bin, i.e. the sum of the sizes of the items contained in the bin.
- The capacity of the bin.
- The size of the item to be allocated.
- The residual space in the current bin.

Similarly, for the two-dimensional strip packing problem, the characteristics used are specific to the problem [82]:

- The difference between the width of the bin and the width of the item.
- The bin height.
- The difference in heights of the bin and opposite bin, i.e. the bin to the right of the bin.
- The difference in heights of the bin and the neighbouring bin.

In addition to problem characteristics, existing heuristics and components of these heuristics can also be incorporated into the heuristics. For example, in the study conducted by Sim and Hart [174] the following bin selection heuristic components were also included in the heuristics:

- Attempts to pack the largest item into the current bin and returns 1 if the attempt is successful and -1 otherwise.
- Attempts to pack the combination of two items with the largest combined size into the current bin. If the attempt is successful 1 is returned, otherwise -1 is returned.

- Attempts to pack the combination of three items with the largest combined size into the current bin. If the attempt is successful 1 is returned, otherwise -1 is returned.
- Attempts to pack the combination of five items with the largest combined size into the current bin. If the attempt is successful 1 is returned, otherwise -1 is returned.
- Attempts to pack the smallest item into the current bin and returns a 1 if the attempt is successful and -1 otherwise.

The generation constructive hyper-heuristic employs a technique to combine the operators with problem characteristics and components of the existing heuristics to create new low-level constructive heuristics. Genetic programming [96] has been used for this purpose for bin packing problems. The operators form the function set and variables representing the problem characteristics and components of existing constructive heuristics form the terminal set for the genetic programming algorithm. Integer constants can also be included in the terminal set and hence can be included in the new evolved heuristics [174]. The function and terminal sets are used to create an initial population, which consists of expression trees representing the low-level heuristics. This initial population is iteratively refined through the processes of evaluation, selection and application of genetic operators to evolve new low-level constructive heuristics for the bin packing problem.

Sim and Hart [174] use a variation of genetic programming, single-node genetic programming (SNGP), to generate constructive heuristics for the one-dimensional bin packing problem. Instead of one implementation of SNGP, a distributed architecture, namely, an island model, is employed in order to explore more than one area of the heuristic space simultaneously. The standard multiplication operator, protected division, relational operators, namely, less than and greater than, and a conditional-branching operator were used. The terminal set comprised of variables representing problem characteristics, components of existing low-level heuristics and integer constants. The problem instances were divided into a training set and a test set. The training set was used by the SNGP island model to create a low-level constructive heuristic, which was then applied to the test set. Hence, the heuristics evolved were reusable. The evolved heuristics performed better than the existing heuristics.

Hyde [82] examined the use of a genetic programming generation constructive hyper-heuristic for the two-dimensional strip packing problem. The heuristics comprised the standard arithmetic operators, protected division and problem characteristics. The evolved heuristics were disposable and performed better than existing constructive heuristics applied separately.

In [82] a hyper-heuristic that creates heuristics for the one- dimensional, two-dimensional and three-dimensional bin packing problems and knapsack problems was investigated. The operators used are the standard addition, subtraction and multiplication operators and the protected-division operator. The problem characteristics catered for all three dimensions and included the x, y and z coordinates of a corner, the volume and value of an item, and three variables that measured the wasted space for each item in terms of each two-dimension combination, i.e. xy, *xz* and *yz*. The evolved heuristics were disposable and performed just as well as the best-performing existing heuristic applied individually.

9.4 Discussion

Hyper-heuristics have proven to be effective at solving bin packing problems. Selection constructive hyper-heuristics have performed better than the existing human-derived heuristics for the one-dimensional and two-dimensional bin packing problems. Similarly, generation constructive hyper-heuristics have created new low-level constructive heuristics that outperform the human-derived existing heuristics for the one-dimensional, two-dimensional and three-dimensional bin packing problems. Furthermore, the time taken to induce these heuristics is less than that required to derive manually. Hence, the use of generation constructive hyper-heuristics provides a means of automating the design of constructive heuristics for bin packing problems.

Multipoint search algorithms have essentially been employed by selection constructive hyper-heuristics. Given the success of single-point search algorithms such as tabu search and variable-neighbourhood search for other problem domains such as examination timetabling [30], it would be interesting to see the effectiveness of single-point search for exploring the heuristic space for bin packing problems. One direction for future research would be to compare the performance of multipoint and single-point algorithms in exploring the heuristic space for bin packing problems.

Given how well selection constructive and generation constructive hyper-heuristics have performed independently for bin packing problems, it would be interesting to examine a hybrid method combining both of these hyper-heuristics. The generation constructive hyper-heuristic would create new heuristics that are contained in the heuristic combinations or condition-action rules generated by the selection constructive hyper-heuristic.

Genetic programming has essentially been used for inducing new constructive heuristics, and has proven to be effective in [82]. However, in [174] single-node genetic programming has been used instead and proved to be more effective than standard genetic programming in inducing effective heuristics for bin packing problems. Furthermore, the heuristics evolved by genetic programming contain redundant code, making the heuristics unreadable. Other techniques, such as grammatical evolution, should be investigated for inducing heuristics.

There has not been sufficient research into the use of perturbative hyper-heuristics for bin packing problems. Both selection constructive and generation constructive hyper-heuristics as well as the hybridization of constructive and perturbative hyper-heuristics should be investigated for bin packing problems.

Chapter 10
Examination Timetabling Problems

10.1 Introduction

Examination timetabling represents one of the earliest and most studied problem domains in hyper-heuristics (HH). Different interesting research issues have been addressed in the literature, from high-level heuristic selection mechanisms and acceptance criteria and designing intelligent low-level heuristics, to fundamental studies on the formal definition of the two search spaces at the two levels. Promising results have been obtained and progress has been made on designing simple and effective approaches across different benchmarks as well as real-world problems.

The examination timetabling problem (ETTP) can be defined as assigning a set of exams, each associated with a number of enrolled students, to a fixed number of slots, subject to satisfying a number of predefined *hard* and *soft* constraints (see Appendix B.4). In the literature on timetabling, a number of benchmark datasets have been introduced in research and competitions, and extensively tested in the last five decades [153]. The main focus has been on a wide range of heuristics and meta-heuristics, and these approaches have been quickly employed in HH. A formal definition of ETTP and the benchmark datasets are presented in Appendix B.4.

10.2 Low-Level Constructive Heuristics for Examination Timetabling Problems

Early research in ETTP focused on constructive heuristics, leading to a rich set of simple heuristics that can be easily used as low-level heuristics *llh* in constructive HH. The basic timetabling problem can be modelled as a graph colouring problem [153], thus graph colouring heuristics have been widely employed in constructive HH, and they perform well across both examination and course timetabling problems [142].

N. Pillay, R. Qu, *Hyper-Heuristics: Theory and Applications*,
Natural Computing Series, https://doi.org/10.1007/978-3-319-96514-7_10

The widely used graph colouring heuristics in constructive HH order exams based on the difficulty of assigning them using the following criteria. The timetable is constructed step by step by assigning the most difficult exams first.

- Largest Degree First (*LD*): exams are ordered in decreasing order by the number of conflicts (i.e. degree) they have with all other exams.
- Saturation Degree First (*SD*): exams are ordered dynamically in ascending order by the number of remaining feasible slots for them during the timetable construction.
- Largest Colour Degree First (*LCD*): exams are ordered in decreasing order by the number of conflicts with those already assigned.
- Largest Enrolment First (*LE*): exams are ordered in decreasing order by the number of students enrolled.
- Largest weighted degree (*LWD*): exams are ordered using *LD*, weighted by the number students involved in the conflict, i.e. taking both exams.
- Highest cost (*HC*): exams are ordered in decreasing order by the cost of violating soft constraints incurred from assigning them to the current timetable.

10.3 Low-Level Perturbative Heuristics for Examination Timetabling Problems

As in the other applications, perturbative HH in ETTP also use simple neighbourhood structures that change a single or multiple variables at each iteration of search. In the literature, a subset of the following pool of *llh* (neighbourhood operators) have usually been employed [34, 132] or extended with other features such as room constraints [54] in perturbative HH.

- *single move*: a single exam is selected and moved to a new feasible slot
- *swap*: slots of two exams are swaped, subject to constraints
- *move n exams*: *n* randomly chosen exams are moved to new slots subject to constraints
- *move whole slot*: an entire randomly selected slot of exams is inserted into a different slot
- *swap slot*: all the exams in two slots are swapped
- *Kempe chain*: a subset of conflicting exams in two slots are swapped
- *constraint based*: exams violating certain soft constraints are selected and moved to a different slot

10.4 Selection Hyper-Heuristics for Examination Timetabling Problems

In timetabling, research has mainly focused on selection HH that configure or choose from the pool of *llh* defined in Sections 10.2 and 10.3.

10.4.1 Selection Perturbative Hyper-Heuristics for Examination Timetabling Problems

Compared to selection constructive HH, less research has been conducted on selection perturbative HH, focusing on heuristic selection and move acceptance criteria [34, 132, 54] at the high-level. The heuristic selection methods used range from the simplest random selection [54] to genetic algorithms [33] to select *llh* from the pool listed in Section 10.3. Move acceptance criteria are listed below. A summary of perturbative HH is provided in Table 10.1.

- *all moves*: all moves are accepted
- *only improving or equal*: only moves to improving or equal solutions are accepted
- *Monte Carlo* or *simulated annealing*: non-improving solutions are accepted probabilistically, and all improving solutions are accepted
- *Great Deluge*: solutions of fitness below a defined threshold are accepted, and the threshold is gradually decreased at each iteration
- *late acceptance*: compared to the solutions in the last few iterations, solutions of better quality are accepted

Table 10.1 Selection perturbative hyper-heuristics in examination timetabling problems

	High-level method	***llh***	**Datasets**
Bilgin et al. [18]	7 heuristic selection, 5 move acceptance	exam and slot selection based on conflicts	Toronto, Yeditepe function optimization
Özcan et al. [132]	choice function with great deluge acceptance	exam selection based on conflicts	Toronto, Yeditepe
Burke et al. [34]	4 heuristic selection, 3 Monte Carlo-based acceptance	4 exam selection based on conflicts	Toronto
Burke et al. [33]	genetic algorithm	23 neighbourhood structures	Toronto
Demeester et al. [54]	simple random tournament selection	4, 3 or 2 move operators, 4 acceptance criteria	Toronto, ITC, KAHO

In a series of research papers [18, 131, 132], a typical iteration of a selection perturbative HH is defined as consisting of two steps, namely heuristic selection and move acceptance at the high level. Combinations of different strategies in the two steps have been analyzed in intensive experimental studies [18]. In [132] reinforcement learning in heuristic selection is investigated in a choice HH with linear decreasing great deluge move criteria. Several factors such as additive and negative adaptation rates and memory length in the learning to reward or punish *llh* have been examined. In [34], among the four different heuristic selection and three Monte Carlo-based acceptance criteria, simulated annealing with reheating showed to be promising when designing selection perturbative HH.

In [54], a simple tournament selection is used at the high level to select the best neighbourhood operators from three groups of *llh* for three post-enrolment and curriculum-based timetabling problems. In-depth analysis has led to observations that are in line with [18], that no single combination of heuristic selection and move acceptance dominates across different problem instances. The authors raised an interesting research question on devising smarter *llh* in selection perturbative HH.

In [33] a new approach is investigated, where a genetic algorithm (GA) is employed to intelligently select appropriate neighbourhoods from a large pool for a variable-neighbourhood search (VNS). The VNS can be seen as configured using a GA, and produced some of the best results at the time. Although 10 out of the 23 neighbourhood operators contributed the most to the significantly improved results in VNS, it is shown that other neighborhood operators are also useful; thus adaptive selection is still needed to perform well across different instances.

10.4.2 Selection Constructive Hyper-Heuristics for Examination Timetabling Problems

Selection constructive HH were the earliest studied for examination timetabling. A summary of these HH is presented in Table 10.2.

Early research in HH focused on indirect encoding in GAs to overcome the limitations of direct encoding and also to improve the generality of construction methods. In [183], indirect encodings of lists of constraint satisfaction strategies (variable and value ordering with the colour degree heuristic) are evolved to construct timetables. At the time the term "hyper-heuristics" had not been specifically used in the paper, however, promising results indicated new directions for developing more general algorithms by evolving combination of heuristics.

In [160] insightful analysis has been conducted on three different fitness measures on variance of solution quality, thus to reflect the overall generality of algorithm performance, i.e. low variance means better generality solving different instances. A steady-state GA with different crossover and mutation is used to evolve a large number of event and slot selection *llh* extended from those in Section 10.2. Using different descriptions of problem states, these *llh* are used to construct timetables for different timetabling problems. Promising results compared to fixed constructive heuristics indicate the synergistic effect from HH, thus encouraging future extensions of research in the area.

A simple and effective graph-based selection hyper-heuristic (GHH) framework is established in [32] to select graph-colouring-based *llh* (as listed in Section 10.2) for constructing timetables. Tabu search is employed at the high level to search heuristic sequences of five graph colouring *llh* and a random ordering strategy. In [151] "two search spaces", namely the heuristic space and solution space, are formally defined within the framework. Interesting issues of both search spaces, such as the upper bound of their sizes, representation and move operators, are discussed. Analysis shows that not all solutions in the solution space can be obtained by the high-level search upon the heuristic space. A simple local search is hybridized within GHH to further improve the solutions, by exploiting solutions not reachable using the high-level search upon the heuristic space. It is proposed that the role of high-level search in this selection constructive GHH is to explore wider regions of solutions, and the local search further exploits solutions obtained in the solution space.

The above GHH framework in [32] searches for heuristic sequences, where appropriate *llh* are used at different stages of solution construction. Extensions of the framework have been investigated in a series of papers. In [152] a two-stage adaptive iterated local search is used at the high level to select those *llh* at the beginning of the heuristic sequence. A local search is then used to improve the whole sequence. Such an approach is designed based on an analysis of a large collection of high-quality heuristic sequences on sample instances. It is observed that those *llh* at the beginning of the heuristic sequences are crucial in constructing high-quality timetables. In [154] an EDA is used to learn those effective *llh* in high quality heuristic sequences based on statistical information collected in a probability vector.

Based on the GHH framework, a type of knowledge-based system named case-based reasoning (CBR) is used to select constructive graph colouring *llh* based on offline knowledge extracted and stored in a case base [40]. Good *llh* used to construct previous high-quality timetables on training instances are stored with their corresponding solution construction scenarios in the case base, and retrieved in similar scenarios when we construct timetables for new instances. Knowledge discovery techniques are used to learn features representing scenarios, and the system is trained based on a large set of randomly generated instances for both course and exam timetabling problems. In [103], *llh* sequences are classified as "good" or "bad" using a neural network model. Only good *llh* sequences are used to construct timetables, in order to significantly reduce the computational expenses of GHH. It has been found that hidden patterns can be found by using a neural network and logistic regression upon the *llh* sequences. This indicates a potential direction of developing other knowledge-based methods for various problems.

Pairwise graph colouring *llh* are simultaneously considered to construct timetables in [7]. The overall "difficulty" of assigning exams is calculated by using fuzzy logic to combine two *llh*, rather than by using a single *llh* as shown in Section 10.2. The combined constructive *llh* has shown to outperform single *llh*, and reduce backtracks (re-assigning exams) in constructing feasible solutions. Although the focus of the research is on combining, rather than selecting, constructive *llh*, the approach could be extended to explore adaptive selection and combination of *llh* using fuzzy models.

In [143], four hierarchical combinations of two graph colouring *llh* are studied. The primary *llh* selects exams and the secondary *llh* breaks the ties while assigning the most difficult exam to the lowest penalty slot. This presents a new way of applying constructive *llh* simultaneously compared with that in [7]. Results obtained are competitive with existing ones in the literature.

The interesting issue of encoding in evolutionary algorithms has been addressed in [148]. Three different encodings of *llh* sequences are compared. It is found that variable-length combination of constructive *llh* performed better. With all three encodings, evolutionary algorithms always converge to the same best *llh* combinations as obtained with individual encodings. The encodings have shown no correlation with the problem characteristics, indicating the generality of this approach across different problem domains.

In [165], a difficulty index is calculated based on four hierarchically ordered lists created using *llh*, to adaptively assign the most difficult exams measured by the combined difficulty. Roulette wheel selection is used to select slots for the chosen exam. This simple and purely constructive approach has shown to be able to produce competitive results compared against the best approaches in the literature.

Table 10.2 Selection constructive HH in ETTP

	High-level method	***llh***	**Dataset**
Terashima-Marín et al. [183]	GA with indirect coding	CSP strategies with colour degree	Toronto
Ross et al. [160]	GA with 3 crossover, 3 mutation operators	16 event selection, 28 slot selection	Toronto, ITC class timetabling
Burke et al. [40]	CBR	LD, CD, SD, LE, LWD	random
Burke et al. [38]	tabu search	LD, CD, SD, LE, LWD, random	Toronto, course timetabling
Qu et al. [152]	adaptive ILS	SD, LWD	Toronto
Qu & Burke [151]	4 local search	LD, LE, CD, SD, LWD	Toronto, course timetabling
Asmuni et al. [7]	fuzzy combination of constructive heuristics	LD, LE, SD	Toronto
Pillay & Banzhaf [143]	4 hierarchical combinations	LD, LWD, SD, LE, HC	Toronto
Li et al. [103]	ANN, logistic regression	LD, CD, SD, LWD	Toronto
Pillay & Rae [148]	EA with three encodings	LD, LWD, SD, LE, HC	Toronto
Sabar et al. [165]	4 hierarchical hybrids of *llh*	LD, LCD, SD, LE	Toronto, ITC07
Qu et al. [154]	EDA heuristics	15 graph colouring	Toronto

10.5 Generation Hyper-Heuristics for Examination Timetabling Problems

In the current literature, there is not much research on generation HH for ETTP. Existing *llh* have been decomposed to create components for high-level generation methods. This requires a good understanding of the specific problems, which cannot be easily transferred across problem domains. These present challenges for raising the generality of efficient generation HH and pose interesting issues for future research in HH.

In [11], in-depth analysis has provided insights into designing effective generation constructive HH while striking a balance between the flexibility of the grammar and the size of the search space. Special initialization is used to generate valid initial trees using the grammar over standard deviation trees. The most widely used graph colouring *llh* in Section 10.2 and slot selection heuristics have been decomposed to create a rich set of components for the generation of sophisticated new heuristics. The approach has shown to outperform other constructive HH as well as some improvement methods due to its ability to evolve new heuristics.

Different high-level methods including genetic programming, genetic algorithms and random generation have been investigated in [146] to evolve new constructive heuristics based on a set of *llh* including problem attributes and period selection heuristics. In two HH, namely arithmetic and hierarchical HH, the interesting issue of two different encodings has also been studied. In genetic programming, trees are used to combine *llh* using function operators, while in genetic algorithms strings of *llh* are used. In hierarchical HH, ties are broken by applying the next attribute in the string. In arithmetic HH, the first event in the ordered list using the generated constructive heuristic is assigned to the period leading to the minimum cost; ties are broken by selecting a random period with the same cost.

Two types of heuristics were examined, arithmetic heuristics and hierarchical heuristics. Arithmetic heuristics were found to perform better than hierarchical heuristics for the examination timetabling problem. It is also interesting that heuristics generated by arithmetic HH are not readable.

10.6 Discussion

ETTP has been investigated in Operational Research and Artificial Intelligence for more than five decades, leading to advanced research in meta-heuristics, exact methods such as mixed integer programming, and hybrid approaches. It also presents one of the earliest-studied applications in HH, and interesting research discoveries have been obtained across several sets of benchmark problems widely tested in the literature.

Based on the rich collection of constructive heuristics and search operators with acceptance criteria, both selection constructive and selection perturbative HH have been developed. Combinations and / or extensions of various heuristics and operators are selected by high-level methods to construct or improve timetable solutions subject to different constraints. The current focus in selection HH for ETTP is on designing a wide range of high-level selection methods, from simple choice functions, random tournament selection and evolutionary algorithms, to various techniques such as fuzzy logic, hybrid ordering strategies, artificial neural networks and case-based reasoning. Research issues associated with developing effective high-level methods include different encodings in evolutionary algorithms, and analysis of the search spaces of heuristics and solutions, leading to interesting findings on the generality and fundamentals of selection HH for ETTP.

There is, however, less research on generation HH for ETTP, compared with that in other applications such as vehicle routing and packing problems. This may be partially due to the lower demand for the generation of new constructive heuristics for ETTP, which have been intensively investigated in meta-heuristics for the last five decades. One potential issue of generation HH is that the generated heuristics are often not readable or not easy to interpret, thus not always reusable for other problems. This is the case not only for ETTP but also for all other problem applications with complex constraints.

Chapter 11
Cross-Domain Hyper-Heuristics

11.1 Introduction

Hyper-heuristics aim to provide heuristic algorithms of a higher level of generality that produce good results for all problems in a domain rather than just for one or two problem instances but poor results for the others. Cross-domain hyper-heuristics extend this scope of generality across domains. These hyper-heuristics aim at producing good results across problems for different domains rather than for one domain and poor results for another domain. This research has essentially focused on solving discrete combinatorial optimization problems. Exclusively selection perturbative hyper-heuristics have been researched to provide cross-domain solutions. This is due to the 2011 CHeSC competition, which challenged the hyper-heuristics community to produce a selection perturbative hyper-heuristic that performs well over six different discrete combinatorial optimization domains [126]. There have been studies that have applied hyper-heuristics to more than one problem domain, however their performance is evaluated separately for each problem domain and not across different problem domains, and hence these are not considered to be cross-domain hyper-heuristics.

Section 11.2 provides an overview of the CHeSC challenge. Selection perturbative hyper-heuristics that have performed well in the challenge and some of the more recent hyper-heuristics applied to solve this problem subsequent to the challenge are outlined in Section 11.3.

11.2 Cross-Domain Heuristic Search Challenge (CHeSC)

In order to promote the use of hyper-heuristics across domains, thereby increasing the level of generality of hyper-heuristics, the Cross-Domain Heuristic Search Challenge (CHeSC 2011) [126] was held in 2011. The HyFlex framework was developed for this purpose, allowing competitors to implement a selection perturbative hyper-

N. Pillay, R. Qu, *Hyper-Heuristics: Theory and Applications*,
Natural Computing Series, https://doi.org/10.1007/978-3-319-96514-7_11

heuristic that would produce good results over all six problem domains. CHeSC 2014 extends the cross-domain challenge by introducing "batching" and allowing multi-threading strategies. The framework provides the following for each domain:

- Methods to create an initial solution.
- Methods to calculate fitness, i.e the objective value.
- Low-level perturbative heuristics.
- Problem instances for the problem domain.

The framework implements the above for the following combinatorial optimization problems:

- Boolean satisfiability
- One-dimensional bin packing
- Permutation flow shop
- Personnel scheduling
- Travelling salesman
- Vehicle routing

The low-level heuristics made available for each of the six domains are generally grouped into four categories [126]:

- Mutational heuristics: Also called perturbation heuristics. These heuristics make small changes to the solution. The operations performed by these heuristics include swapping, changing, adding or deleting solution components.
- Ruin-recreate heuristics: These destruction-construction heuristics delete some components of a solution, and then reconstruct the solution using problem specific low-level construction heuristics.
- Local search heuristics: Also called hill-climbing heuristics. These heuristics make small changes to a solution iteratively. The new solution is only accepted if it has fitness equal to or better than the original solution to which the heuristic was applied. These heuristics guarantee that the new solution will be at least as good as the original solution.
- Crossover: This heuristic is applied to two selected solutions and produces a single offspring.

The performance of a hyper-heuristic is assessed over the problem instances for all six problem domains in comparison with of the performance of other hyper-heuristics also applied to these problems. The eight top-performing hyper-heuristics are determined based on the objective value of the solution produced by the hyper-heuristic. These hyper-heuristics are assigned a rank, with a higher rank indicating a better performance. The remaining hyper-heuristics are not assigned a rank. These ranks are then summed to indicate the performance of each hyper-heuristic over all the problem instances and different problem domains.

11.3 Approaches Employed by the Hyper-Heuristics

This section provides an overview of the top five finalists in the CHeSC challenge in Section 11.3.1 and recent approaches that produced good results in Section 11.3.2.

11.3.1 Finalists of CHeSC 2011

The winning approach in the challenge was AdapHH. AdapHH [117, 120] maintains an adaptive dynamic heuristic set for heuristic selection. At the end of a set number of iterations, the heuristics are evaluated according to their performance during these iterations. A performance metric measuring heuristic performance in terms of how well the heuristic performs, i.e. the best solutions produced, and its speed, is used for this purpose. A performance index calculated by the performance metric is used to calculate a quality index for each heuristic. The heuristics with a quality index below the average quality index of all the heuristics are omitted from the set for a number of iterations, referred to as the tabu duration. If the tabu duration of a heuristic is increased due to consecutive omissions, and exceeds a maximum threshold, the heuristic may be permanently removed from the heuristic set. After a set number of iterations, extreme heuristic removal may occur, where heuristics not performing well on all the iterations are removed from the set. A heuristic is selected from the set using a selection probability, which is a function of its performance and time taken when it is applied to the problem. The study also introduces the concept of relay hybridization to identify a pair of heuristics that can be applied consecutively. Adaptive iteration limited list-based threshold accepting is introduced as the move acceptance to accept better moves. In addition it also accepts moves that result in worse solutions when compared to a solution of a previous iteration, not the last iteration.

In [80, 81] a variable-neighbourhood search selection perturbative hyper-heuristic (VNS-TW) is developed for the six combinatorial optimization problems. This hyper-heuristic performs four steps, namely, shaking, local search, environmental selection and periodic adjustment, iteratively. During shaking a low-level heuristic from the mutation and ruin-create categories of heuristics is selected and applied to a base solution. On the first iteration, the base solution is randomly selected. Local search involves iteratively applying heuristics from the local search heuristic category, and terminates when there is no improvement for a set number of iterations. The number of iterations is a parameter, initially set to a maximum value and adjusted during the periodic adjustment step. A dynamically sized population of good solutions is maintained. During environmental selection, the new solution produced at the end of local search replaces a worse solution in the population, and a new base solution is selected from the population for the next iteration. Periodic adjustment is invoked whenever the approach has been running for a set time budget, to adjust the population size and the number of iterations for which there is no improvement in the local search. This approach was placed second in the CHeSC 2011 challenge.

The approach ML using reinforcement learning employed by Larose [98] was placed third in the challenge. ML iterates between a diversification cycle, an intensification cycle and move acceptance. During the diversification cycle, heuristics are selected from the mutation and recreate and no-op (has no effect) heuristics. The intensification cycle selects local search operators. Reinforcement learning as employed in [113] is used to select the heuristics to apply. The heuristics are chosen based on their performance during previous iterations. A weight matrix is maintained to keep track of heuristic performance, and is adjusted according to heuristic performance. The solution produced after the intensification and diversification phases is accepted if it is an improvement on the solution of the current iteration, or there has been no improved solution for a set number of iterations.

PHunter, the selection perturbative hyper-heuristic employed by Chan et al. [53], takes an analogy from pearl hunting. Pearl hunting involves a diver diving to retrieve pearls and resurfacing. When the diver resurfaces the next dive will be moved to a new area to retrieve the pearls from this area. The phases of diving and moving to a new area are seen as analogous to the processes of intensification and diversification in search. During the "move" or diversification phase, a heuristic is selected from any category except the local search heuristics. If the resulting solution has an objective value that is within a specified threshold, this process continues. During a "dive" or intensification, local search heuristics are applied. Two types of dives are performed, namely, snorkelling and deep dive. Snorkelling involves applying a short sequence of hill climbers with a low search depth. Deep dive applies a long sequence of hill climbers with a high search depth iteratively until there is an improvement in the objective value. Snorkelling is first performed to obtain a set of solutions, which are ranked. Deep dive is then applied to a subset of the best solutions resulting from snorkelling. PHunter also works in different modes. For example, if during snorkelling and deep dive the heuristics produce the same solution, this is referred to as "shallow water". This results in only snorkelling being performed with a simplified sequence of heuristics. Offline learning is performed to determine the different modes. PHunter was placed fourth in the challenge.

The fifth finalist was an evolutionary programming hyper-heuristic (EPH) [112]. It employs co-evolution, where a population of solutions and a population of heuristic sequences are evolved at the same time. The solutions are initially created using a construction heuristic for the problem domain, e.g. first-fit for the one-dimensional bin-packing problem. The initial population of heuristic sequences is randomly created. Each heuristic sequence comprises a perturbation segment followed by a local search segment. The perturbation segment comprises one or two heuristics from the mutation, crossover or ruin-create categories. The local search segment contains local search heuristics. Each local search heuristic is applied either once, or iteratively using variable-neighbourhood descent where only solutions with an improved objective value are accepted. Each heuristic sequence is evaluated by applying it to a randomly selected solution from the population. The resulting solution replaces the worst solution in the population if its objective value is better than at least one solution in the population, and its objective value is different to that of the solutions already in the population. Tournament selection and the mutation operator are used to create successive generations of heuristic sequences.

Subsequent to the competition, there have been a number of initiatives to improve the results obtained by the finalists for the challenge. The following section provides an overview of these recent attempts.

11.3.2 Recent Approaches

As the field advances, some recent selection perturbative hyper-heuristics have performed better than most of the finalists described in the previous section. In [168] the low-level perturbative heuristics are modelled as a tree. Monte Carlo search is then applied to this tree to select the heuristic to apply. The approach includes a memory mechanism which stores a population of solutions that are produced by applying a heuristic. Each heuristic is applied to a solution randomly selected from memory. If the solution is better than the current solution, it replaces this solution in memory. The move acceptance component accepts all improving solutions as well as worse solutions according to a preset probability. This hyper-heuristic outperformed the five finalists in the competition.

In [92] heuristic sequences consisting of low-level heuristics are first created. These are then modelled as hidden Markov models, with each heuristic corresponding to a hidden state in the Markov model and assigned a probability of being selected. The move acceptance criterion is a binary decision that either accepts all moves or accepts a solution if it is better than that produced by the previous heuristic sequence evaluated. This selection perturbative hyper-heuristic outperformed the five finalists in the challenge.

An iterated multi-stage hyper-heuristic is employed in [93]. In this study two hyper-heuristics are applied in cycles. One of the hyper-heuristics, the stage2 hyper-heuristic, is used to determine the effectiveness of the low-level heuristics by using a greedy selection method. Based on their performance, a score is assigned to each low-level heuristic. The subset of the best-performing heuristics is used by the second hyper-heuristic, the stage1 hyper-heuristic. In the stage1 hyper-heuristic, roulette wheel selection is used to select a heuristic and threshold acceptance is used to decide whether to accept the heuristic or not. Threshold acceptance accepts all improving solutions and worse solutions that are not worse than the objective value of the best solution obtained thus far by a threshold. The stage1 hyper-heuristic is applied first with all the low-level heuristics scored equally. It also incorporates relay hybridization, which pairs heuristics together and applies them consecutively. Each low-level heuristic is applied for a set duration. The stage2 hyper-heuristic is then applied to reduce the set of low-level heuristics. In cycling between the hyper-heuristics, no further improvement objective value is used as the criterion to switch between the two hyper-heuristics. This approach also outperformed the five finalists presented in the previous section.

The authors in [9] employ a tensor-based hybrid move acceptance hyper-heuristic for the cross-domain challenge. The hyper-heuristic performs five phases, namely, noise elimination, tensor construction, tensor factorization, tensor analysis and hybrid move acceptance. Noise elimination determines the set of low-level heuristics to use, i.e. it eliminates poorly performing heuristics. A tensor is constructed using these heuristics and factorized to determine which pairs of heuristics perform well together. Tensor analysis involves dividing the heuristics into two partitions, with each partition using a different move acceptance method. These form two selection perturbative hyper-heuristics, which are applied in a round-robin manner for a set duration. This process is applied iteratively to solve the problem at hand. This hyper-heuristic produced better results than four of the finalists, namely, ML, VNS-TW, PHunter and EPH.

11.4 Discussion

This chapter presents cross-domain hyper-heuristics, i.e. hyper-heuristics that perform well over problem instances from different combinatorial optimization domains. This research has essentially focused on selection perturbative hyper-heuristics as a result of the CHeSC 2011 challenge and the public availability of the HyFlex framework; see Appendix A.1. If we examine the top-performing selection perturbative hyper-heuristics certain trends are apparent that appear to contribute to the success of these cross-domain hyper-heuristics. A majority of these hyper-heuristics include a mechanism for reducing the set of low-level heuristics at some point. In AdapHH [120] a dynamic heuristic list is maintained from which poorly performing heuristics are eliminated for a set duration or completely. Similarly, in the study conducted by Kheiri and Özcan [93] a second hyper-heuristic is used to determine the set of low-level heuristics to use, and the noise elimination phase in [9] determines the set of low-level heuristics to use. Furthermore, the move acceptance methods employed also accept worsening moves in addition to moves to an improvement in the objective value, which enables the search to escape local optima. In two of the best performing hyper-heuristics, the effectiveness of relay hybridization has been illustrated and warrants further investigation. A thorough theoretical analysis of these hyper-heuristics, including fitness landscape analysis, needs to be conducted to better understand their performance.

All the research in this area has involved selection perturbative hyper-heuristics. There has also been some research into the automated generation of selection perturbative cross-domain hyper-heuristics [124, 167, 169]. These studies will be discussed in Chapter 13 on automated design of hyper-heuristics. However, there is a lack of research into selection constructive hyper-heuristics or generation hyper-heuristics to solve cross-domain problems. Future research is needed to address this.

Part III
Past, Present and Future

Chapter 12
Advances in Hyper-Heuristics

12.1 Introduction

The previous chapters have introduced the four types of hyper-heuristics, presented the theoretical foundations and examined various applications of hyper-heuristics. This chapter provides an overview of some advanced topics and recent trends in hyper-heuristics, namely, hybrid hyper-heuristics, hyper-heuristics for automated design, automated design of hyper-heuristics and hyper-heuristics for continuous optimization.

12.2 Hybrid Hyper-Heuristics

Hybrid hyper-heuristics refers to combining two or more types of hyper-heuristics, namely, selection constructive, selection perturbative, generation constructive and generation perturbative, to solve a problem. In most of the studies conducted in this area, two hyper-heuristics are combined. In [79], a generation constructive hyper-heuristic employing genetic programming is used to create new constructive heuristics. A genetic algorithm selection constructive hyper-heuristic is used to determine the most effective sequence of the new heuristics to solve the problem. The hybrid hyper-heuristic is applied to the capacitated vehicle routing problem and outperformed the constructive heuristics generally used to solve the problem.

Li et al. [102] use a hybrid hyper-heuristic to minimize the total weighted tardiness in the intercell scheduling problem considering transportation capacity. A generation perturbative hyper-heuristic is used to evolve rules to improve a solution. The rules to apply in solving the problem are selected using a selection perturbative hyper-heuristic. The generation constructive hyper-heuristic employs genetic programming to evolve the rules. The selection constructive hyper-heuristic uses a genetic algorithm combined with local search to select which rules to apply. The hybrid hyper-heuristic outperformed the human-created rules for solving the problem.

N. Pillay, R. Qu, *Hyper-Heuristics: Theory and Applications*,
Natural Computing Series, https://doi.org/10.1007/978-3-319-96514-7_12

Sim and Hart [175] apply a hybrid hyper-heuristic to solve the vehicle routing problem. A generation constructive hyper-heuristic employing genetic programming is used to evolve constructive low-level heuristics, which are used to create a population of solutions. A selection perturbative hyper-heuristic using a memetic algorithm is then used to improve the population of initial solutions. The results produced by the hyper-heuristic were found to be competitive with the known optimum for the vehicle routing problem instances tested.

Miranda et al. [116] employ a hybrid hyper-heuristic, supported by a knowledge base, for the automated design of particle swarm optimization (PSO) algorithms. A generation perturbative hyper-heuristic using grammar-based genetic programming is used to produce the algorithms. The grammar specifies the procedures, e.g. the strategy to use to initialize the swarm, and parameters, e.g. mutation probability, that can be combined by the genetic programming algorithm to produce PSO algorithms. A selection perturbative hyper-heuristic is used, based on a case-based reasoning system, to select which evolved PSO algorithm to apply to solve the problem at hand. In case-based reasoning, a knowledge base stores characteristics of the problem (i.e. fitness landscape of the problem) and the corresponding evolved algorithm for solving the problem. When solving a new problem, the characteristics of the problem are compared to those stored in the knowledge base to retrieve the algorithm applied to the most similar previous problems to the new problem. Euclidean distance is used to measure the similarities between the problems. The hybrid hyper-heuristic performed better than standard PSO algorithms in solving 60 optimization problems.

12.3 Hyper-Heuristics for Automated Design

Hyper-heuristics have proven to be effective at automating the design of various machine learning and search techniques. Selection pertubative and generation perturbative hyper-heuristics have been used for this purpose. A summary of some of the research conducted in this area is presented in this section. Selection perturbative hyper-heuristics have been used to select parameter values, operators, selection methods and metaheuristics as the low-level heuristics. Generation perturbative hyper-heuristics have been used to generate operators, rules and algorithms.

In the study conducted by Hong et al. [78], a generation pertubative hyper-heuristic employing genetic programming is used to evolve mutation operators for evolutionary programming for solving function optimization problems. The operators were induced using a training set, and were found to generalize well when used in evolutionary programming to solve a test set of problems. Evolutionary programming with the evolved mutation operators was found to perform better than when used with human designed operators.

Branke et al. [20] have used a generation perturbative hyper-heuristic to generate dispatching rules for dynamic stochastic job shop scheduling. The hyper-heuristic employed an evolutionary algorithm to explore the space of dispatching rules. Three representations for the dispatching rules were compared, namely, expression trees, neural networks and linear combinations. The tree representation showed to be the most effective.

Lourenço et al. [107] employed a grammatical evolution selection perturbative hyper-heuristic to automate the design of an evolutionary algorithm to solve the knapsack problem. The hyper-heuristic selects the control model for the evolutionary algorithm, e.g. generational, the operators and their probabilities and the selection method and associated parameter values, e.g. tournament size. The performance of the evolutionary algorithms produced by the hyper-heuristic were found to be competitive with manually designed evolutionary algorithms.

Barros et al. [15] used a genetic programming generation pertubative hyper-heuristic, HEAD-DT, to induce decision tree induction algorithms for data classification. The hyper-heuristic was tested on 20 binary and multiclass classification problem instances. Its performance was compared to C4.5 and CART, which are state-of-the-art decision tree induction algorithms. The induction algorithms produced by HEAD-DT outperformed both C4.5 and CART in terms of predictive accuracy.

Falcão et al. [58] implemented a selection perturbative hyper-heuristic to select meta-heuristics, and their parameters, to use at each state of solving a problem. In this case, the low-level heuristics are the meta-heuristics and parameter values. The hyper-heuristic was used to solve a scheduling problem involving the allocation of tasks with limited resources. Reinforcement learning was used to perform these selections. The hyper-heuristic outperformed a multi-agent approach applied to solve this problem.

Maashi et al. [108] employed a selection pertubative hyper-heuristic to select a multi-objective evolutionary algorithm to be applied at each point in solving a multi-objective optimization problem. The low-level heuristics include NSGA-II, SPEA2 and MOGA. The hyper-heuristic used a choice function for heuristic selection and great deluge or late acceptance for move acceptance. On solving two problem domains, namely, the Walking Fish Group test suite and vehicle crashworthiness design problem, the hyper-heuristic performed competitively with NSGA-II.

In [64], a selection pertubative multi-objective evolutionary algorithm was used to design a stacked neural network. The design decisions made by the hyper-heuristic include the number of neural networks that must be included in the stack, the weights of the outputs for the neural networks, and the number of hidden neurons for each neural network. The multi-objective evolutionary algorithm hyper-heuristic combined NSGA-II with the quasi-Newton optimization algorithm. The hyper-heuristic was tested on a real-world problem, namely, the modelling of polyacrylamide-based multicomponent hydrogel synthesis. The stacked neural networks designed by the hyper-heuristic outperformed the human-designed stacked neural networks.

12.4 Automated Design of Hyper-Heuristics

While hyper-heuristics have been shown to be effective for the automated design of various machine learning techniques and meta-heuristics, a recent research direction in the field is the automated design of hyper-heuristics. This section reports on some of the studies focusing on this. A majority of this research has focused on the design of selection perturbative hyper-heuristics.

Choong et al. [46] used reinforcement learning, namely, Q-learning, to design a selection perturbative hyper-heuristic. Q-learning is used to design the heuristic selection component and move acceptance component of the hyper-heuristic by selecting from six heuristic selection methods and five move acceptance techniques. The selection perturbative hyper-heuristic is an iterated local search incorporating the selected components. The iterated local search also includes an intensification and diversification phase. The approach produced competitive results on the domains in the CHeSC 2011 cross-domain challenge.

In the study conducted by Sabar et al. [169], gene expression programming was used to induce the move acceptance component of a selection perturbative hyper-heuristic. The evolved move acceptance components comprise arithmetic operators and terminal values representing the quality of the previous and current solution, the current iteration, and the number of iterations completed. The heuristic selection component uses a credit reward mechanism that takes into consideration the previous performance of the heuristic. This mechanism is used together with a dynamic multi-armed bandit mechanism to determine which heuristic to use. The latter takes into consideration the reward determined by the credit reward mechanism and the number of times the heuristic was applied to decide which heuristic to use. The approach performed competitively, outperforming some manually designed hyper-heuristics when applied to the vehicle routing problem, examination timetabling problem and domains in the CHeSC 2011 cross-domain challenge.

The research above was extended in [167], in which both the heuristic selection and move acceptance components of the selection perturbative hyper-heuristic are induced using gene expression programming. Each element of the population is composed of two components, one representing the heuristic selection and one the move acceptance. Each component comprises arithmetic and logical operators combined with terminal values. The terminal set for the heuristic component contains different values representing the performance of the low-level perturbative heuristic, while the terminal set for the move acceptance component consists of different values pertaining to the quality of the previous and current solution and the current and total number of iterations. In order to maintain diversity, the generated selection perturbative hyper-heuristic maintains a memory mechanism containing high-performing diverse solutions, which are updated throughout the application of the hyper-heuristic. This approach was applied to the CHeSC 2011 cross-domain challenge, and outperformed the finalists of the challenge.

Fontoura et al. [62] applied a similar approach using grammatical evolution to generate a heuristic selection component and move acceptance component for a selection perturbative hyper-heuristic for solving the protein structure prediction problem. The generated heuristic selection components comprise arithmetic operators combined with terminal values representing the previous performance of the low-level perturbative heuristics. Similarly, the induced move acceptance components comprise arithmetic operators and terminal values. These values represent the quality of previous and current solutions and the number of iterations performed. The approach produced the best results for seven of the 11 protein structure predication problem instances.

In [8], the authors used an apprenticeship-learning-based hyper-heuristic to generate a selection perturbative hyper-heuristic to solve the vehicle routing problem. This hyper-heuristic produces a set of classifiers based on the performance of an expert selection perturbative hyper-heuristic, i.e. the winner of the CHeSC 2011 challenge AdapHH, in solving the vehicle routing problem. Each classifier is essentially a production rule, where the action is the low-level perturbative heuristic to apply, the move acceptance criterion or a parameter value of a low-level heuristic, and the condition represents the state of the search. The selection perturbative hyper-heuristics produced by the apprenticeship-learning-based hyper-heuristic were found to outperform the expert hyper-heuristics and other selection perturbative hyper-heuristics in solving the vehicle routing problem.

A similar approach was taken in [187], where a time delay neural network instead of apprenticeship learning is employed by the hyper-heuristic to generate the classifiers. The parameters of the neural network are determined using Taguchi orthogonal arrays. In this study a single classifier is generated. The attributes for the classifier are the difference in the objective value and the generated solution from one iteration to the next, and the class is the low-level perturbative heuristic to apply. Hence, in this study, just the heuristic selection component of the selection perturbative hyper-heuristic is induced. The move acceptance criterion is accept all moves. The induced selection perturbative hyper-heuristics outperformed the expert hyper-heuristic.

12.5 Continuous Optimization

Initially hyper-heuristics aimed at solving discrete optimization problems, and hence a majority of the research has focused on this. However, recently the effectiveness of hyper-heuristics in solving continuous optimization problems has been illustrated. This section provides an overview of some the research conducted in this area.

The selection perburbative hyper-heuristic applied by Maashi et al. [108] described in the previous section to hybridize multi-objective evolutionary algorithms has proven to be effective at solving continuous optimization problems. Similarly, the genetic programming generation perturbative hyper-heuristic employed by Hong et al. [78] to evolve mutation operators for the evolutionary programming algorithm produced good results for function optimization problems.

Segredo et al. [172] employed a selection perturbative hyper-heuristic to hybridize differential evolution and a genetic algorithm to solve the problem instances of the generalization-based contest on global optimization. In this case the low-level perturbative heuristics are the differential evolution and genetic algorithm. The hyper-heuristic assigns scores to the low-level heuristics based on previous performance. The hyper-heuristic probabilistically selects whether to apply the best-performing heuristic or a randomly selected heuristic. The hybrids created by the hyper-heuristic performed better than differential evolution and the genetic algorithm applied individually, and won two of the three contest prizes.

Walker and Keedwell [190] used a selection perturbative hyper-heuristic to select a sequence of low-level perturbative heuristics to solve multi-objective continuous optimization problems. The hyper-heuristic uses a hidden Markov model to determine the sequence of low-level perturbative heuristics, and performed competitively with, and for some problems instances better than, existing algorithms when applied to the problem instances from the DTLZ benchmark set.

12.6 Discussion

This chapter provides an overview of some advanced topics in hyper-heuristics. As the field is constantly growing, this list is by no means complete. Hybrid hyper-heuristics combining the strengths of two types of hyper-heuristics have been illustrated. This has illustrated the potential of hybridizing hyper-heuristics however, at most two hyper-heuristics have been hybridized. Hence, further hybridization including the combination of more than two hyper-heuristics should be investigated. Furthermore, the reasons why these hybridizations work well and how best to combine the different types of hyper-heuristics need to be examined by studying the theoretical aspects of the hybridization, such as movement through the search spaces, correlations between search in the heuristic and solution spaces, fitness landscapes and many more.

There has been a fair amount of research into the use of hyper-heuristics for the automated design of machine learning and meta-heuristic techniques. Hyper-heuristics have proven to be effective for this purpose. The design decisions made by hyper-heuristics range from parameter selection, selection of operators and operator probabilities, and hybridization of approaches, to the generation of new operators, rules and algorithms. The use of hyper-heuristics for automated design reduces the man-hours required, thereby enabling the researcher to focus on other aspects such as the problem domain. The aim in using hyper-heuristics for automated design is not to produce results that are competitive with state-of-the-art techniques, but to automate the design process, leading to results at least as good as that achieved by manual design. However, from the overview provided in the previous section, it is evident that the automated design produces approaches that outperform the manually designed approaches. Hyper-heuristics have employed various techniques for automated design, with evolutionary algorithms being the most popular. Future research in this area should also focus on investigating which techniques are the most appropriate for which design decisions.

A more recent area of research has been the automated design of hyper-heuristics, the effectiveness of which has been illustrated in the overview of some of the studies provided in this chapter. However, the research has only focused on selection perturbative hyper-heuristics, and the majority of the applications have been to the CHeSC 2011 cross-domain challenge. The automated design of the remaining three types of hyper-heuristics, namely, selection constructive, generation constructive and generation perturbative, needs to be investigated. Furthermore, the automated design of hybrid hyper-heuristics should also be examined. Hyper-heuristics could be used to generate hybrid hyper-heuristics, with the low-level heuristics being the hyper-heuristics.

While the majority of the research in the area of hyper-heuristics to this date has focused on applying hyper-heuristics to discrete optimization problems, there has been recent research investigating their use to solve continuous optimization problems. The majority of this research has involved using a hyper-heuristic to automate the design of the approach. Investigations are needed into using the four types of hyper-heuristics to directly solve continuous optimization problems. The use of cross-domain hyper-heuristics to solve continuous optimization problems and cross-domain hyper-heuristics to solve both continuous and discrete optimization problems also needs to be researched.

Chapter 13
Conclusions and Future Research Directions

Recent research advances have been made in different types of hyper-heuristics (HH), namely selection HH and generation HH, employing both constructive and perturbative low-level heuristics (*llh*). Among the four types of HH, selection HH (Chapters 2, 3) received more research attention than generation HH (Chapters 4, 5). This may be due to the research challenges in developing genetic programming and grammatical evolution, which are the main high-level techniques used in generation HH. These include the issue of bloating, which leads to the problem of readability and interpretability [14]. Among most of the generation HH, the newly generated *llh* have thus rarely been reused on new instances or problems. This presents challenges but interesting research directions for further investigations.

A large number of high-level methods have been investigated in HH. These include single-point and multiple-point meta-heuristics including local search and evolutionary algorithms, and various techniques including case-based reasoning [16, 37, 40], choice function [17, 52, 89, 132], fuzzy logic [6, 7], grammatical evolution [57, 146, 164], genetic programming [84, 96, 104, 82, 174, 193], Markov chains [92, 91], Monte Carlo [168, 34], rules [4], simple random method [17, 54], and hybridizations between them. Most of these have been studied in both selection and generation HH for examination timetabling problems (Chapter 10). The investigations of genetic programming have been mostly conducted in generation HH for vehicle routing problems (Chapter 7) but not in nurse rostering problems (Chapter 8). Investigations of these various techniques across different problem domains of diverse problem characteristics can lead to further research findings and strengthen fundamental discoveries on landscapes of high-level and low-level search spaces in HH (see Chapter 6).

A good range of *llh* have been employed; some are problem specific while others are commonly used across different applications. In the case of perturbative *llh*, these can be combined together with acceptance criteria. These research findings on different *llh* for different problem domains provide good ground for further in-depth

N. Pillay, R. Qu, *Hyper-Heuristics: Theory and Applications*,
Natural Computing Series, https://doi.org/10.1007/978-3-319-96514-7_13

investigations in terms of the generality and efficiency of HH. For example, different groups of *llh* with different execution speed and the number of changes to problem solutions in selection HH [119] have been investigated to gain insights into their contributions to the generality of HH performance. It is proposed that features of *llh* in relation to the generality in HH should be analysed by particular mechanisms to adaptively manage and select *llh* and to design general HH. The synergy between constructive and perturbative *llh* should also be examined to further improve the efficiency of HH.

A new formal definition for a general HH of different types is presented in Chapter 6, based on an existing definition in [151] for selection HH with constructive *llh*. Within the general HH, two optimization problems have been defined at two levels, respectively, each associated with an objective function, namely $f(s)$ for the low-level search space of problem solutions s, and $F(h)$ for the high-level search space of heuristics h. A mapping function M associates the search within the two spaces, i.e. M: $f(s) \rightarrow F(h)$. Some fundamental issues on landscape studies and analysis of the features of the search spaces have been discussed. Further investigations and understanding of the search spaces can facilitate the design of more effective HH. Other fundamental studies, such as runtime analysis of selection perturbative HH, have been conducted in [100]. It is shown that online reinforcement learning in HH does not outperform that of HH with a fixed distribution of *llh* operators. Such investigations into other types of HH may reveal further interesting findings; thus underpinning the fundamentals and theory of general HH across more problems.

A good range of applications has been studied in recent HH research, including vehicle routing in Chapter 7, nurse rostering in Chapter 8, packing problems in Chapter 9, examination timetabling in Chapter 10, as well as real-world combinatorial optimization problems [30]. This presents a nice and diverse range of representative applications. Compared to the other applications, more results have been obtained on generation HH for packing problems. At the time of writing this book, there is a lack of research on generation HH for nurse rostering, which, compared to the other applications, involves more types of constraints. For all the applications covered in this book, there exist well established benchmark datasets in the existing literature; thus comparison studies can be conducted, leading to interesting observations in both HH and meta-heuristic communities.

Although HH aims to increase the generality of search algorithms in solving different problems and instances, in the existing literature the majority of HH approaches have been tested on a single, and some on several specific problem domains, each evaluated separately against a particular objective function. The generality of the HH approaches has not been measured against certain standard or unified criteria across different problem domains. In a recent study, an initial attempt has been made to establish the measurement of four different levels of generality when assessing the generality of HH approaches [147], compared against specific evaluations for different problems. More such measurements in future HH developments will underpin research towards designing general algorithms across a range of different combinatorial optimization problems.

Since the inception of the field, there have been various advances in HH research. One such area is hybrid HH (Section 12.2). While there have been initial studies in this area, there is a need for further investigations such as the hybridization of more than two hyper-heuristics. HH have successfully been used for automated design (Section 12.3). The design decisions that have been automated using HH range from parameter tuning to creating new operators. An emerging area is the automated design of HH, which has contributed to reducing the man-hours involved in HH design [125]. The majority of HH research has focused on solving discrete optimization problems; however, more recently this been extended to continuous optimization problems as well (Section 12.5). Additional emerging areas include using HH to solve multi-objective optimization problems [108] and dynamic optimization problems [95].

In HH, domain specific knowledge can be considered by the *llh* for the problem under consideration, leaving the high-level search problem independent. That is, the general search is handled at the high level, isolated from the details of constraints and structure of solutions for the specific problem. In all the existing research, constraint handling has been conducted at the low-level solution space, by either discarding infeasible solutions constructed or generated, or by employing targeted operators that explore only feasible solutions. Investigations on effective constraint handling techniques, in conjunction with their effect on the connectivity of both search spaces, can enhance the performance HH for highly complex and constrained problems.

In HH approaches, both online and offline learning have been used to improve the efficiency of search upon *llh*. These include the offline learning of rules by using artificial neural networks [4] to construct nurse rostering solutions, and learning and storing constructive heuristics in a case-based reasoning system to construct timetables at different stages [37]. Online learning is usually conducted by adaptively adjusting the rewards or scores of *llh* based on the solutions generated. Examples include choice function [89] and reinforcement learning [113, 132]. There is, however, no extensive study on different types of learning in HH. Such investigations, employing for example machine learning techniques, could open new interesting research directions and further enhance the generality of HH approaches. For example, in [103], artificial neural networks have been trained offline to identify potentially high-quality nurse rostering solutions. During the problem solving on new instances, only those potential rosters of high quality are selected and evaluated, to reduce the large amount of computational time spent unnecessarily evaluating all roster solutions. Such a mechanism is highly effective in solving those complex and constrained problems, which is the case in HH, where a large amount of computational time is spent evaluating the generated solutions at the low level. Other existing research in machine learning, for example on fitness estimation in evolutionary algorithms [86], could also be explored within HH in future research.

Appendix A
HyFlex and EvoHyp

Along with the recent advances in hyper-heuristics to raise the generality of search algorithms across different problem domains, frameworks and toolkits have been developed in the literature. In this book, details of a widely used framework HyFlex and a toolkit EvoHyp are provided in Appendix A. These two software suites are open-source, and can be accessed and used to develop hyper-heuristic approaches.

HyFlex is a framework that researchers and practitioners can use to implement hyper-heuristics. This framework provides common software interfaces as well as problem specific components for developing cross-domain general search methodologies. HyFlex has been widely used for nurse rostering (Chapter 8), vehicle routing (Chapter 7), and across different domains (Chapter 11). Details are provided in Appendix A.1.

While HyFlex provides a framework within which users can implement selection perturbative hyper-heuristics to solve the six problems, EvoHyp provides a toolkit for developing evolutionary algorithm hyper-heuristics to solve problems. The problem domain must be implemented by the user. EvoHyp allows the user to implement a genetic algorithm selection constructive, a genetic algorithm selection perturbative or a genetic programming generation constructive hyper-heuristic to solve a particular problem. EvoHyp also provides distributed versions of each of the hyper-heuristics. Appendix A.2 provides an overview of EvoHyp.

A companion site[1] with links to HyFlex and EvoHyp will be updated with other software and sources in hyper-heuristics in the future.

[1] https://sites.google.com/view/hyper-heuristicstheoryandapps

N. Pillay, R. Qu, *Hyper-Heuristics: Theory and Applications*,
Natural Computing Series, https://doi.org/10.1007/978-3-319-96514-7

A.1 HyFlex

In the 2011 Cross-Domain Heuristic Search Challenge (CHeSC 2011) [29, 126], the hyper-heuristics community was challenged to develop general hyper-heuristics within a framework called HyFlex, written in Java, for solving the following six discrete combinatorial optimization problems. For each problem domain, 10 training instances have been derived from benchmark datasets at different sources. The last two problem domains are used as the hidden problem domains in CHeSC2011.

- Boolean satisfiability: In this Maximum Satisfiability (MAX-SAT) problem, an assignment of the Boolean variables in a formula needs to be determined so that the whole formula is satisfied, i.e. evaluated to be true. The 10 instances with 250-744 variables have been derived from the Maxsat Evaluation 2009 benchmark datasets [2] and the two SAT competitions [3] [4].
- One-dimensional bin packing: For the classic one-dimensional bin packing problems (see Appendix B.1), an alternative fitness function has been employed. The 10 instances with 160-5,000 items and bins of capacity 150-1,000 have been derived from the European Special Interest Group on cutting and packing benchmark datasets [5].
- Permutation flow shop: In the permutation flow shop problems, a number of given jobs need to be scheduled and processed on a set of machines, subject to the ordering constraints, i.e. processed in a predefined order of machines. The objective is usually to minimize the makespan (i.e. the completion time of the last finished job). The 10 instances in HyFlex with 100 or 200 jobs to be scheduled to 10 or 20 machines are derived from the flow shop benchmark datasets [6].
- Personnel scheduling: Personnel scheduling at hospital wards worldwide usually involves a large number of constraints (see Appendix B.2). In Hyflex, problem instances are derived from [83] and the staff rostering benchmark datasets [7]. The problems involve scheduling 12-51 staff to 3-12 shifts spanning 26-42 days.
- Travelling salesman problem: TSP represents one of the most studied problems in combinatorial optimization. The benchmark instances from the TSPLIB [158] [8] have been used in HyFlex, with problem sizes from 299 to 13,509 cities.
- Vehicle routing problem: Based on the Solomon dataset and Gehring and Homberger datasets (see Appendix B.3), 10 instances of capacitated vehicle routing problems with time windows have been derived and included. These include instances of 20 or 250 vehicles with capacity of 200 or 1,000, and 1,000 customers of three types, namely located randomly, clustered in groups and clustered randomly.

---

[2] http://www.maxsat.udl.cat/

[3] http://www.cril.univ-artois.fr/SAT07

[4] http://www.cril.univ-artois.fr/SAT09/

[5] http://paginas.fe.up.pt/~esicup/

[6] http://mistic.heig-vd.ch/taillard/

[7] http://www.schedulingbenchmarks.org/

[8] http://elib.zib.de/pub/mp-testdata/tsp/tsplib/tsp/index.html

Within the HyFlex framework, problem specific components include four categories of low-level perturbative heuristics and problem instances for the above six combinatorial optimization problems. These categories of perturbative heuristics are:

- Mutational heuristics - Equivalent to perturbation heuristics and operators, which make small changes to the solution variables based on the gains of the evaluation function. The operations performed by these heuristics include swapping, changing, adding or deleting solution components.
- Ruin-recreate heuristics - These destruction-construction heuristics randomly reassign values to a proportion of variables in the solution, i.e. remove some components of a solution, and then reconstruct it using problem specific low-level construction heuristics.
- Local search heuristics - These hill-climbing heuristics iteratively make small changes to randomly selected variables in a solution. The acceptance is first-improvement, i.e. the first solution of a better or equal fitness is accepted.
- Crossover - These standard one-point or two-point crossover operators are applied to two selected solutions to produce a single offspring.

HyFlex is built with various general mechanisms and problem specific components for developing selection perturbative hyper-heuristics. The general mechanisms include methods to create an initial solution, and methods to calculate fitness, i.e the objective value. Details of these mechanisms are provided in Chapter 11. In CHeSC2014, HyFlex was extended to include "batching" with multi-threading strategies.

Analysis of the winning approaches in CHeSC2011 are presented in Chapter 11. More details of the results, resources and documentation are available on the CHeSC website [9].

[9] http://www.asap.cs.nott.ac.uk/external/chesc2011/

A.2 EvoHyp

EvoHyp is a Java toolkit to implement evolutionary algorithm hyper-heuristics. EvoHyp provides four libraries: *GenAlg*, *GenProg*, *DistrGenAlg* and *DistrGenProg* [145]. An overview of these libraries is given below.

A.2.1 GenAlg

GenAlg implements a generational genetic algorithm to provide a genetic algorithm selection hyper-heuristic. This can be implemented as a selection constructive or a selection perturbative hyper-heuristic. In the case of selection constructive hyper-heuristics, the combination will be used to create an initial solution to the problem; for selection perturbative hyper-heuristics, the heuristic combination is used to improve an initial solution. Tournament selection is used to choose parents to which the mutation and crossover operators are applied, to create the offspring of each generation. The genetic algorithm terminates when a maximum number of generations has been reached.

In *GenAlg*, the user has to:

- Specify the parameter values, e.g. population size and number of generations, for the genetic algorithm.
- Specify the characters representing the low-level heuristics.
- Define the problem domain in terms of:
  - Implementation of the low-level heuristics.
  - A method that will apply the heuristic combination produced by the hyper-heuristic and calculate the fitness of the heuristic combination.
  - A method that determines whether one heuristic combination is fitter than another.

A.2.2 GenProg

GenProg employs a genetic programming algorithm to create new low-level heuristics. These heuristics can be an arithmetic function or an arithmetic rule. In the case of arithmetic functions, the hyper-heuristic combines arithmetic operators with characters representing attributes of the problem domain to create new heuristics. These attributes can also be existing low-level heuristics or components thereof. In the case of arithmetic rules, the problem attributes are combined with arithmetic operators as well as an *if-then-else* operator. Each element of the population is a parse tree representing an arithmetic function or rule. The grow method [96] is used to create the initial population. The generational algorithm is used to evolve the initial population over a number of generations. As in the case of *GenAlg* the user has to provide a function to calculate the fitness of each parse tree in the population as part of the problem domain implementation. Tournament selection is used to choose parents to which mutation and crossover are applied to create the population of each generation. The algorithm terminates when a maximum number of generations is reached. The user has to:

- Specify the parameters for the genetic programming algorithm.
- Specify the characters representing the problem attributes.
- Define the problem domain in terms of:
  - A method that uses the arithmetic function or rule to create a solution to the problem. The method must calculate the fitness of the arithmetic rule or function based on the solution it produces.
  - A method that determines which of two arithmetic functions or rules is fitter than the other.

A.2.3 Distributed GenAlg and GenProg

EvoHyp includes libraries for the distributed versions of *GenAlg* and *GenProg*, namely, *DistrGenAlg* and *DistrGenProg*, respectively. The aim of these libraries is to reduce the runtimes associated with implementation of these evolutionary algorithm hyper-heuristics. *DistrGenAlg* distributes the implementation of the genetic algorithm over a multicore architecture. Similarly, *DistrGenProg* distributes the genetic programming algorithm over a multicore architecture. In both instances, this is achieved by dividing the population size by the number of cores available to obtain n subpopulations. During initial population generation and application of the genetic operators, each subpopulation is created and evaluated on a different core.

A.2.4 Accessing EvoHyp

EvoHyp can be accessed following the website[10]. There are currently two versions, 1.0 and 1.1. The difference in the versions is that EvoHyp1.1 displays the solution created by the best-performing heuristic combination or heuristic on each generation, and 1.0 does not provide such details.

[10] https://sites.google.com/view/hyper-heuristicstheoryandapps

Appendix B
Combinatorial Optimization Problems and Benchmarks

In hyper-heuristics, the most tested combinatorial optimization problems (COPs) in operational research have been employed to demonstrate the generality of these algorithms. This section presents the definition, problem model and constraints for these COPs. Details of these are also provided online [1]; they thus also serve as a collection of benchmark COPs for scientific comparison and analysis in both hyper-heuristics and meta-heuristics in artificial intelligence and optimization.

[1] https://sites.google.com/view/hyper-heuristicstheoryandapps

N. Pillay, R. Qu, *Hyper-Heuristics: Theory and Applications*,
Natural Computing Series, https://doi.org/10.1007/978-3-319-96514-7

B.1 Packing Problems

Packing problems essentially involve packing items into bins or containers, so that a minimum number of bins or containers is used [13]. These packing problems can be one dimensional (1D), two dimensional (2D) or three dimensional (3D).

B.1.1 One-Dimensional Bin Packing

The 1D bin packing problem requires a set of items of varying sizes to be packed into bins so as to meet the following conditions:

- The capacity of each bin is not exceeded.
- A minimum number of bins is used.

A formal definition of the one-dimensional bin packing problem is provided in Definition B.1

Definition B.1. Given a set of bins of capacity C, a given set of n items of different sizes $S = s_1, ..., s_n$ must be packed into a minimum number of bins m subject to the capacity constraint, i.e. the fullness of each bin f_i does not exceed the bin capacity, i.e. $f_i \leq C$, for $i = 1, ..., m$.

There are two versions of the problem, namely, offline and online. In the offline version of the problem, the sizes of the items are known before packing, while in the online version the sizes are only known at the time of packing [59, 171]. In most versions of the problem, the capacity of each bin C is the same; however, in some versions the capacity of each bin is different.

B.1.2 Two-Dimensional Bin Packing

The 2D bin packing problem is a variation of the 1D bin packing problem, where the bin size is not defined in terms of capacity, but in terms of the width and height of each bin. Furthermore, each item to be packed is specified in terms of a width and height. The aim is to pack the items into the bins subject to the following constraints:

- The items in each bin must not exceed the dimensions of the bin.
- The items in a bin must not overlap.
- A minimum number of bins must be used.

A formal definition of the two-dimensional bin packing problem is presented in Definition B.2.

Definition B.2. In 2D bin packing, a set of n items with their widths $W = w_1,...,w_n$ and heights $H = h_1,...,h_n$ must be packed into a minimum number of bins m so that the fullness of each bin $f_i, i = 1,...,m$, does not exceed the width and height of each bin, and the items do not not overlap in the bin.

There are a number of variations of the 2D bin packing problem such as the orthogonal packing without rotation [13], 2D irregular packing [105] and the 2D strip packing problems [82].

B.1.3 Three-Dimensional Bin Packing

The 3D bin packing problem is an extension of 2D bin packing, where the size of each bin is defined in terms of a width, height and depth. Similarly, the size of each item is also expressed in terms of a width, height and depth. The items must be packed in the bins subject to the following conditions:

- The items in each bin must not exceed the dimensions, width, height and depth, of the bin.
- The items in a bin must not overlap.
- A minimum number of bins must be used.

Definition B.3 provides a formal definition of the three-dimensional bin packing problem.

Definition B.3. In 3D bin packing, a set of n items with width $W = w_1,...,w_n$, height $H = h_1,...,h_n$ and depth $D = d_1,...,d_n$ must be packed into a minimum number of bins m so that the fullness of each bin $f_i, i = 1,...,m$, does not exceed the width, height and depth of each bin, and the items do not not overlap in the bin.

As in the case of the 2D problem, based on how the items are permitted to be packed, variations of this problem also exist [109].

B.1.4 Packing Benchmark Sets

The benchmark sets used for 1D online packing, 2D and 3D packing have been used in the specific studies and a number of these sets have been generated as part of the study. However, for the 1D bin packing problem, there are two benchmark sets that are commonly used for the offline version of the problem, namely, the Falkenauer benchmark set [59] and the Scholl benchmark set [171].

The Falkenauer dataset [2] consists of two classes of problems, namely, the uniform class and the triplets class. In the uniform class the problem instances require

---

[2] http://people.brunel.ac.uk/∼mastjjb/jeb/orlib/binpackinfo.html

items of size uniformly distributed in the range 20 to 100 to be packed into bins with capacity 150. The problem instances in the triplets class requires items with sizes in the range 25 to 50 to be packed into bins with capacity 100.

The Scholl dataset [3] comprises of three datasets, namely, dataset 1, dataset 2 and dataset 3. The datasets consist of 720, 480 and 10 problem instances respectively. Dataset 3 contains hard problem instances. The datasets and instances comprising each set differ in terms of the range of the sizes of the items and the capacities of the bin.

B.2 Nurse Rostering Problem

Definition B.4. The Nurse Rostering Problem (NRP) involves constructing roster solutions by assigning a set of nurses N_n with different skills to a set of shifts S_s of different types over a scheduling period D_d, satisfying a set of constraints C_c. The objective is to minimize the violations of constraints C_c in the generated roster solutions.

Due to the diverse range of legislations across different countries, a large variety of hard and soft constraints have been modelled in the NRP literature. Hard constraints must be satisfied. Those different constraints and requirements that may be violated are defined as soft constraints. Violations of soft constraints are often used to measure the quality of the rosters, and used by solution methods as the evaluation function. The objective is to minimize the violations of soft constraints while satisfying the hard constraints, examples of which are presented in Table B.1.

Table B.1 Examples of hard and soft constraints in NRP

Hard Constraints
Coverage: all shifts must be assigned during the scheduling period
One nurse can take at most one shift per day
Soft Constraints
Maximum / minimum number of shift assignments during the scheduling period
Maximum / minimum consecutive working days during the scheduling period
Maximum / minimum free days / weekends during the scheduling period
Maximum / minimum free days between shifts during the scheduling period
Personal preferences

Several benchmark datasets have emerged along with the extensive research on NRP in the last five decades for research comparisons. The most common constraints have been included in the benchmark datasets.

[3] https://www2.wiwi.uni-jena.de/Entscheidung/binpp/index.htm

B.2.1 The 2010 International Nurse Rostering Competition

The 2010 International Nurse Rostering Competition [76] (INRC2010) aims to close the gap between theory and practice by introducing problem models with higher complexity. Based on developments in the existing research, INRC2010 has promoted a range of new approaches in NRP research.

The scheduling period consists of four weeks, with four or five shifts per day. The two hard constraints are presented in Table B.1. Each nurse has a contract, which defines the legal constraints as presented in Table B.1[76, 17].

With the same problem model, different computational times (seconds, minutes and hours) and sizes are defined in three tracks of instances, namely *Sprint*, *Middle Distance* and *Long Distance* to reflect different challenges in practice. All instances are provided in XML and text format, along with the benchmarking of computational time for different machines and platforms. The problem data and competition rules can be found at the competition site [4].

B.2.2 The UK Benchmark Nurse Rostering Dataset

One of the early NRP benchmark datasets includes 52 instances derived from three wards in a major UK hospital [4]. In the problem, 20-30 nurses need to be assigned to two shift types, *day* (early and late) and *night* shifts. Nurses of three grades are contracted to work either days or nights in one week but not both. The aim is to produce weekly roster schedules with evenly distributed unsatisfied requests and unpopular shifts.

To address this highly constrained problem, various preferences and historical assignments of shifts to nurses have been used to build a collection of 411 valid weekly shift patterns of day and night shifts, with their associated costs. The problem complexity is thus captured in these patterns, the desirability of which for the nurses is indicated by the costs. These valid pre-processed shift patterns are use to construct weekly rosters in a range of hyper-heuristics (Chapter 8) and meta-heuristics in the literature.

B.2.3 The Nottingham Benchmark Nurse Rostering Dataset

A benchmark NRP web site [5] has been established at the University of Nottingham to collect and maintain a wide range of NRP problems worldwide. Countries involved include Belgium, Canada, Finland, Japan, Netherlands, and UK, etc. The problem description is provided in an XML format, to formulate flexibly the features

[4] http://www.kuleuven-kortrijk.be/nrpcompetition

[5] http://www.schedulingbenchmarks.org/

and complex and diverse constraints. The best solutions reported in the literature are also provided with the lower bound obtained, and updated with the corresponding references.

For NRP with highly complex constraints, collections of such problem instances with the corresponding solutions in a consistent format are valuable to promote research in the hyper-heuristics, meta-heuristics and optimization communities.

B.3 Vehicle Routing Problems

The basic vehicle routing problem (VRP) involves scheduling a set of closed routes R_1, ..., R_m beginning and ending at a depot v_0, each taken by a vehicle k_1, ..., k_m, to serve an ordered list of tasks v_1, ..., v_n (customer locations). The objective is to minimize the total distance of R_r. In some problem models, the number of vehicles is also minimized. Definition 7 provides a definition of VRP.

Definition B.5. The basic VRP is usually modelled as a network $G = (V, A)$, where $V = \{v_0, v_1, ..., v_n\}$ is the set of nodes representing the depot v_0 and customers v_1, ..., v_n, and $A = \{(v_i,v_j), v_i,v_j \in V, i \neq j\}$ represents the set of links between tasks / customers v_i and v_j; each (v_i,v_j) is associated with a cost (distance) $d_{i,j}$.

A large number of VRP variants have been defined and investigated in the literature to evaluate different algorithms and techniques [97, 186]. The most studied variants include the following features or constraints:

- VRP with time window constraints (VRPTW) [22]: each customer task $v_1...v_n$ is associated with a time window (a_i, d_i), a_i and d_i is the arrival and departure times, respectively, within which v_i must be served.
- VRP with capacity constraints (CVRP) [68]: each vehicle k_1, ..., k_m is associated with a certain capacity c_1, ..., c_m, which must be satisfied when serving all customers on each route R_1, ..., R_m.
- VRP with different customer tasks, either pickup and delivery (VRPPD): the remaining capacity varies depending on the task type.
- Open VRP where vehicles may not need to return to the depot (OVRP): each route R_1, ..., R_m starts at v_0 but does not need to finish at v_0.
- VRP with dynamic customer requests (DVRP) [136, 170, 159]: new customer tasks v_i may arrive within a time horizon of [0, T], and are added to VRP during the scheduling period. The objective is to minimize the rejected customer tasks.

These variants may be combined or extended to form new VRP variants with multiple constraints and additional features (i.e. uncertainties). CVRPTW is one of the most studied variants due to its common occurrence in real-world applications.

B.3.1 Vehicle Routing Problem Benchmark Datasets

Over the years, a number of datasets have been established, providing benchmarking testbeds for research in meta-heuristics and evolutionary computation [97, 186, 68, 22, 72]. In addition to the problem size, different aspects of constraints and features have been addressed, leading to a wide variety of problem instances to evaluate the robustness of algorithms and techniques. Table B.2 presents a summary of the features of the benchmark datasets used in hyper-heuristics reviewed in this book. This also represents a set of most-used representative datasets in meta-heuristics and evolutionary computation.

Table B.2 Benchmark VRP datasets

Datasets	**Problem Features**
Christofides Beasley [47]	Seven CVRP instances with 50-200 customers randomly distributed *R* or grouped in clusters *C* in a Cartesian coordinate system
Solomon [176]	56 CVRPTW instances with 100 customers, in six classes with different density of time windows, long and short scheduling horizons, and three types of customers: *R*, *C* and *RC*, served by 7-19 vehicles. The objective is to minimize the number of vehicles and travel distance
Fisher [61]	12 instances with 25-199 customers, most of which are centralized around the depot, serviced by 3-16 vehicles of uniform capacity
Homberger -Gehring [77]	56 Solomon instances, and five groups, each of 60 VRPTW instances, with 200, 400, 600, 800 and 1,000 customers of three types: *R*, *C* and *RC*. The objective is to minimize the number of vehicles and total travel distance
CHeSC2011	Five CVRPTW instances taken from the *Solomon* and *Gehring & Homberger* datasets, respectively, with 100 or 250 customers of three types: *R*, *C* and *RC*, served by 20 or 250 vehicles with capacity of 200 or 1,000 (see Appendix A.1 and Chapter 11). Also available at the SINTEF Transportation Optimization Portal https://www.sintef.no/vrptw
Saint-Guillain [170]	Five classes of 15 DVRPTW instances with 100 stochastic customers, categorized by low to high degrees of dynamism (ratio of dynamic requests). Each class has different distributions of early or late requests in three stages of the scheduling horizon

B.4 Examination Timetabling Problems

Definition B.6. Examination timetabling problems can be defined as requiring a set of exams $E = \{e_1, e_2, ..., e_e\}$ to be assigned to a limited number of ordered timeslots (time periods) $T = \{t_1, ..., t_t\}$ and into rooms of certain capacity $C = \{C_1, ..., C_t\}$ in timeslot t, subject to a set of constraints [153].

In the timetabling literature, the complexities arise from the large variety of constraints in different institutions. In general, constraints are categorized into two types:

- *Hard Constraints* cannot be violated under any circumstances. A timetable that satisfies all of the hard constraints is said to be *feasible*.
- *Soft Constraints* are desirable but may be violated when it is impossible to satisfy all of them. Soft constraints vary from one institution to another in both types and importance [153]. The quality of timetables is usually measured based on to what extent the soft constraints are violated in the timetables.

Due to the large variety of examination timetabling problems investigated in the literature, it would be neither practical nor beneficial to present a comprehensive list of all the hard and soft constraints. Some benchmark examination timetabling problems have emerged in the last five decades [153]. Table B.3 lists some of the key hard and soft constraints. A detailed list of hard and soft constraints can be found in [153].

Table B.3 Example of common hard and soft constraints in examination timetabling problems

Hard Constraints	**Definition**
Conflict: No exams with common students assigned simultaneously	If exams e_i and e_j with D_{ij} common students assigned to timeslots t_i and t_j, then $t_i \neq t_j$, $\forall e_i, e_j \in E, e_i \neq e_j, D_{ij} > 0$
Capacity: The total room capacity for all exams scheduled in timeslot t needs to be sufficient	For all exams e_i scheduled to timeslot t of a total capacity C_t, each e_i with s_i students, $\Sigma_{e_i \in E}\, s_i \leq C_t$, $t_i = t$, $t \in T$
Soft Constraints	
Spread conflicting exams as much as possible throughout T Schedule all large exams as early as possible	

B.4.1 Exam Timetabling Benchmark Datasets

The Toronto dataset was first introduced in [43], and has been widely tested in the last 30 years [153]. It consists of 13 instances from different institutions, among which 11 have been more heavily investigated due to inconsistencies in the other two instances. A detailed discussion of this dataset (and the difficulties caused by different instances circulating under the same name) was given in [153]. The constraints in the problems can be outlined as follows:

- Hard constraint: no conflicting exams (with common students) should be scheduled in the same timeslot.
- Soft constraint: to spread conflicting exams across the timetable.

Table B.4 presents the characteristics of the 11 problems in the dataset. The "conflict density" provides the density of elements C_{ij} with value of 1 in the conflict matrix, where element $C_{ij} = 1$ if events i and j conflict, $C_{ij} = 0$ otherwise. The penalty of the timetable generated is the sum of costs per student, where costs w_i, $i \in \{0, 1,$

2, 3, 4}, is weighted by the number of timeslots with two conflicting exams. More details can be found online [6] and in [153].

Table B.4 Characteristics of the benchmark exam timetabling problems

Instances	car91	car92 I	ear83 I	hec92 I	kfu93	lse91	sta83 I	tre92	ute92	uta93 I	yok83 I
Exams	682	543	190	81	461	381	139	261	184	622	181
Timeslots	35	32	24	18	20	18	13	23	10	35	21
Students	16,925	18,419	1,125	2,823	5,349	2,726	611	4,360	2,750	21,266	941
Conflict Density	0.13	0.14	0.27	0.42	0.6	0.6	0.14	0.18	0.8	0.13	0.29

[6] http://www.cs.nott.ac.uk/~pszrq/data.htm

References

1. Aamodt, A., Plaza, E.: Case-based reasoning: Foundational issues, methodological variations, and system approaches. Artificial Intelligence **1**, 39–52 (1994)
2. Abdullah, S., Ahmadi, S., Burke, E., Dror, M.: Investigating Ahuja-Orlin's large neighbourhood search for examination timetabling. OR Spectrum **29**(2), 351–372 (2007)
3. Ahmed, L., Özcan, E., Kheiri, A.: Solving high school timetabling problems worldwide using selection hyper-heuristics. Expert Systems with Applications **42**, 5463–5471 (2015)
4. Aickelin, U., Li, J.: An estimation of distribution algorithm for nurse scheduling. Annals of Operations Research **155**(4), 289–309 (2007)
5. Aron, R., Chana, I., Abraham, A.: A hyper-heuristic approach for resource provisioning-based scheduling in gird environment. Journal of Supercomputing **71**, 1427–1450 (2015)
6. Asmuni, H., Burke, E., Garibaldi, J.: Fuzzy multiple ordering criteria for examination timetabling. In: Burke E.K. and Trick M. (eds.) Selected Papers from the 5th International Conference on the Practice and Theory of Automated Timetabling, pp. 334–353. Lecture Notes in Computer Science 3616 (2005)
7. Asmuni, H., Burke, E., Garibaldi, J., McCollum, B., Parkes, A.: An investigation of fuzzy multiple heuristic orderings in the construction of university examination timetables. Computers & Operations Research **36**(4), 4981–1001 (2009)
8. Asta, S., Özcan, E.: An apprenticeship learning hyper-heuristic for vehicle routing in Hyflex pp. 1474–1481 (2014)
9. Asta, S., Özcan, E.: A tensor-based selection hyper-heuristic for cross-domain heuristic search. Information Sciences **299**, 412–432 (2015)
10. Bader-El-Den, M., Poli, R.: Generating SAT local-search heuristics using a GP hyper-heuristic framework. In: Artificial Evolution: International Conference on Artificial Evolution, pp. 37–49. Springer (2008)
11. Bader-El-Den, M., Poli, R., Fatima, S.: Evolving timetabling heuristics using a grammar-based genetic programming hyper-heuristic framework. Memetic Computing **1**, 205–219 (2009)
12. Bai, R., Burke, E., Kendall, G., Li, J., McCollum, B.: A hybrid evolutionary approach to the nurse rostering problem. IEEE Transactions on Evolutionary Computation **14**(4), 580–590 (2011)
13. Bansal, N., Correa, J.R., Kenyon, C., Sviridenko, M.: Bin packing in multiple dimensions: Inapproximability results and approximation schemes. Mathematics of Operations Research **31**(1), 31–49 (2006)

N. Pillay, R. Qu, *Hyper-Heuristics: Theory and Applications*,
Natural Computing Series, https://doi.org/10.1007/978-3-319-96514-7

14. Banzhaf, W., Nordin, P., Keller, R.E., Francone, F.D.: Genetic Programming: An Introduction On the Automatic Evolution of Computer Programs and Its Applications. Morgan Kaufmann Publishers (1998)
15. Barros, R.C., Basgalupp, M.P., de Carvalho, A.C., Freitas, A.A.: A hyper-heuristic evolutionary algorithm for automatically designing decision-tree algorithms. In: Proceedings of the 14th Annual Conference on Genetic and Evolutionary Computation (GECCO'12), pp. 1237–1244 (2012)
16. Beddoe, G., Petrovic, S.: Selecting and weighting features using a genetic algorithm in a case-based reasoning approach to personnel rostering. European Journal of Operational Research **175**(2), 649–671 (2006)
17. Bilgin, B., Demeester, P., Misir, M., Vancroonenburg, W., Berghe, G.: One hyper-heuristic approach to two timetabling problems in health care. Journal of Heuristics **18**(3), 401–434 (2012)
18. Bilgin, B., Özcan, E., Korkmaz, E.: An experimental study on hyper-heuristics and exam timetabling. In: Proceedings of the International Conference on the Practice and Theory of Automated Timetabling (PATAT 2006), pp. 394–412 (2006)
19. Blum, C., Roli, A.: Metaheuristics in combinatorial optimization: Overview and conceptual comparison. ACM Computing Surveys **35**(3), 268–308 (2003)
20. Branke, J., Hildebrandt, T., Scholz-Reiter, B.: Hyper-heuristic evolution of dispatching rules: A comparison of rule representations. Evolutionary Computation **23**(2), 249–277 (2015)
21. Branke, J., Nguyen, S., Pickardt, C., Zhang, M.: Automated design of production scheduling heuristics: A review. IEEE Transactions on Evolutionary Computation **20**(1), 110–124 (2016)
22. Bräysy, O., Gendreau, M.: Vehicle routing problem with time windows, part ii: Metaheuristics. Transportation Science **39**, 119–139 (2005)
23. Brucker, P., Burke, E., Curtois, T., Qu, R., Berghe, G.: A shift sequence based approach for nurse scheduling and a new benchmark dataset. Journal of Heuristics **16**(4), 559–573 (2010)
24. Bull, L.: Applications of Learning Classifier Systems, *Studies in Fuzziness and Soft Computing*, vol. 150, chap. Learning Classifier Systems: A Brief Introduction, pp. 1–12. Springer (2004)
25. Burke, E., Bykov, Y., Newall, J., Petrovic, S.: A time-predefined local search approach to exam timetabling problems. IIE Transactions **36**(6), 509–528 (2004)
26. Burke, E., Causmaecker, P.D., Berghe, G., Landeghem, H.: The state of the art of nurse rostering. Journal of Scheduling **7**(6), 441–499 (2004)
27. Burke, E., Dror, M., Petrovic, S., Qu, R.: Hybrid graph heuristics with a hyper-heuristic approach to exam timetabling problems. In: B. Golden, S. Raghavan, E. Wasil (eds.) The Next Wave in Computing, Optimizatin and Decision Technologies - Conference Volume of the 9th Informs Computing Society Conference, 79-91 (2005)
28. Burke, E., Eckersley, A., McCollum, B., Petrovic, S., Qu, R.: Hybrid variable neighbourhood approaches to university exam timetabling. European Journal of Operational Research (206), 46–53 (2015)
29. Burke, E., Gendreau, M., Hyde, M., Kendall, G., McCollum, B., Ochoa, G., Parkes, A.J., Petrovic, S.: The cross-domain heuristic search challenge - an international research competition. In: Springer, Proc. Fifth International Conference on Learning and Intelligent Optimization (LION5), vol. 6683, pp. 631–634. Lecture Notes in Computer Science (2011)
30. Burke, E., Gendreau, M., Hyde, M., Kendall, G., Ochoa, G., Özcan, E.: Hyper-heuristics: A survey of the state of the art. Journal of Operational Research Society **64**, 1695–1724 (2013)
31. Burke, E., Hyde, M., Kendall, G., Ochoa, G., Özcan, E., Woodward, J.: A classification of hyper-heuristic approaches. In: Handbook of metaheuristics, pp. 449–468 (2010)
32. Burke, E., Hyde, M., Kendall, G., Woodward, J.: Automatic heuristic generaiton with genetic programming: Evolving a jack-of-all-trades or a master of one. In: Proceedings of the 9th Annual Conference on Genetic and Evolutionary Computation, vol. 2, pp. 1559–1565 (2007)
33. Burke, E., Hyde, M., Kendall, G., Woodward, J.: A genetic programming hyper-heuristic approach for evolving two dimensional strip packing heuristics. IEEE Transactions on Evolutionary Computation pp. 942–958 (2010)

34. Burke, E., Kendall, G., Misir, M., Özcan, E.: Monte Carlo hyper-heuristics for examination. Annals of Operations Research **196**(1), 73–90 (2012)
35. Burke, E., Kendall, G., Newall, J., Hart, E., Ross, P., Schulenburg, S.: Hyper-heuristics: An emerging direction in modern search technology. In: Handbook of metaheuristics, pp. 457–474 (2009)
36. Burke, E., Kendall, G., Soubeiga, E.: A tabu-search hyperheuristic for timetabling and rostering. Journal of Heuristics **9**, 451–470 (2003)
37. Burke, E., MacCarthy, B., Petrovic, S., Qu, R.: Knowledge discovery in a hyper-heurisitc for course timetabling using case-based reasoning. In: Lecture Notes in Computer Science, vol. 2740, pp. 90–103. Springer (2002)
38. Burke, E., McCollum, B., Meisels, A., Petrovic, S., Qu, R.: A graph-based hyper-heuristic for educational timetabling problems. European Journal of Operational Research **176**, 177–192 (2007)
39. Burke, E., Newall, J.: Solving examination timetabling problems through adaptation of heuristic orderings. Annals of operations Research **129**, 107–134 (2004)
40. Burke, E., Petrovic, S., Qu, R.: Case-based heuristic selection for timetabling problems. Journal of Scheduling **9**(2), 115–132 (2006)
41. Burke, E., Qu, R., Soghier, A.: Adaptive selection of heuristics within a grasp for exam timetabling problems. In: Proceedings of the 4th Multidisciplinary International Scheduling Conference: Theory and Applications (MISTA 2009), pp. 409–423 (2009)
42. Caramia, M., DellOlmo, P., Italiano, G.: New algorithms for examination timetabling. In: Dell' Olmo, Naher, S., Wagner, D. (eds.) Algorithm Engineering., pp. 230–241. Lecture Notes in Computer Science 1982 (2001)
43. Carter, M., Laporte, G., Lee, S.: Examination timetabling: Algorithmic strategies and applications. Journal of Operational Research Society **47**, 373–383 (1996)
44. Causmaecker, P.D., Berghe, G.: A categorization of nurse rostering problems. Journal of Scheduling **14**, 3–16 (2011)
45. Chen, P., Kendall, G., Berghe, G.: An ant based hyper-heuristic for the travelling tournament problem. In: IEEE Symposium on Computational Intelligence in Scheduling (SCIS'07), p. doi: 10.1109/SCIS.2007.367665 (2007)
46. Choong, S.S., Wong, L.P., Lim, C.P.: Automatic design of hyper-heuristic based on reinforcement learning. Information Sciences (2018). DOI doi:10.1016/j.ins.2018.01.005
47. Christofides, N., Beasley, J.: The period routing problem. Networks **14**(2), 237–256 (1984)
48. Clarke, G., Wright, J.: Scheduling of vehicles from a central depot to a number of delivery points. Operations Research **12**(4), 568–581 (1964)
49. Contreras-Bolton, C., Parada, V.: Automatic design of algorithms for optimization problems. In: Proceedings of the 2015 Latin-America Congress on Computaitonal Intelligence (LA-CCI2015) (2015)
50. Cordeau, J., Gendreau, M., Hertz, A., Laporte, G., Sormany, J.: New heuristics for the vehicle routing problem. In: Logistics Systems: Design and Optimization, pp. 279–297 (2005)
51. Cowling, P., Kendall, G., Soubeiga, E.: A hyperheuristic approach to scheduling a sales summit. In: Practice and Theory of Automated Timetabling III, LNCS 2079, pp. 176–190 (2001)
52. Cowling, P., Kendall, G., Soubeiga, E.: Hyper-heuristics: A robust optimization method applied to nurse scheduling pp. 851–860 (2002)
53. C.Y. Chan Fan Xue, W.I., Cheung, C.: A hyper-heuristic inspired by pearl hunting. http://www.asap.cs.nott.ac.uk/external/chesc2011/entries/xue-chesc.pdf (2011)
54. Demeester, P., Bilgin, B., Causmaecker, P.D., Berghe, G.: A hyperheuristic approach to examination timetabling problems benchmarks and a new problem from practice. Journal of Scheduling **15**(1), 83–103 (2012)
55. Drake, J.: Crossover control in selection hyper-heuristics: Case studies using MKP and Hyflex. Ph.D. thesis, School of Computer Science (2014)
56. Drake, J., Hyde, M., Ibrahim, K., Özcan, E.: A genetic programming hyper-heuristic for the multidimensional knapsack problem. Kybernetes **43**(9/10), 1500–1511 (2014)

57. Drake, J., Killis, N., Özcan, E.: Generation of VNS components with grammatical evolution for vehicle routing. In: Proceedings of the 16th European Conference on Genetic Programming (EuroGP'13), pp. 25–36 (2013)
58. Falcao, D., Madureira, A., Pereira, I.: Q-learning based hyper-heuristic for scheduling system self-paramterization. In: Proceedings of the 2015 10th Iberian Conference on Information Systems and Technologies (2015). DOI doi: 10.1109/CISTI.2015.7170394
59. Falkenauer, E.: A hybrid grouping genetic algorithm for bin packing. Journal of Heuristics **2**(1), 5–30 (1996)
60. Ferreira, A., Pozo, A., Gonçalves, R.: An ant colony based hyper-heuristic approach for the set covering problem. Advances in Distributed Computing and Artificial Intelligence Journal (2015)
61. Fisher, M.: Optimal solution of vehicle routing problems using minimum k-trees. Operations Research **42**(4), 626–642 (1994)
62. Fontoura, V.D., Pozo, A.T., Santana, R.: Automated design of hyper-heuristic components to solve the psp problem with hp model. In: Proceedings of the 2017 IEEE Congress on Evolutionary Computation, pp. 1848–1855 (2017)
63. Fukunaga, A.: Automated discovery of local search heuristics for satisfiability testing. Evolutionary Computation **16**(1), 31–61 (2008)
64. Furtuna, R., Curteanu, S., Leon, F.: Multi-objective optimization of a stacked neural network using an evolutionary hyper-heuristic. Applied Soft Computing **12**(1), 133–144 (2012)
65. Garey, M., Johnson, D.: Computers and Intractability: A Guide to the Theory of NP-Completeness. W. H. Freeman & Co., New York, NY, USA (1979)
66. Garrido, P., Riff, M.: Dvrp: a hard dynamic combinatorial optimisation problem tackled by an evolutionary hyper-heuristic. Journal of Heuristics **16**(6), 795–834 (2010)
67. Gaspero, L.D., Schaerf, A.: Tabu search techniques for examination timetabling. In: Burke E.K. and Erben W. (eds.): Selected Papers from the 3rd International Conference on the Practice and Theory of Automated Timetabling, pp. 104–117. Lecture Notes in Computer Science 2079 (2001)
68. Gendreau, M., Laporte, G., Potvin, J.Y.: Chapter 6. metaheuristics for the capacitated VRP, year = 2002. In: T. P., V. D. (eds.) The Vehicle Routing Problem, SIAM Monographs on Discrete Mathematics and Applications, Vol. 9, pp. 129–154. Springer
69. Gillett, B., Miller, L.: A heuristic algorithm for the vehicle dispatch problem. Operation Research **22**, 340–349 (1974)
70. Glover, F., Laguna, M.: Tabu Search. Kluwer Academic Publishers (1997)
71. Goldberg, D., Korb, B., Deb, K.: Messy genetic algorithms: Motivation, analysis and first results. Complex Systems **3**, 493–530 (1989)
72. Golden, B., Raghavan, S., Wasil, E.A.: The Vehicle Routing Problem: Latest Advances and New Challenges. Springer Science & Business Media, Vol 43 (2008)
73. Han, L., Kendall, G.: Guided operators for a hyper-heuristic genetic algorithm. In: A1 2003: Advances in Artificial Intelligence, pp. 807–820 (2003)
74. Hansen, P., Mladenovic, N.: Variable neighbourhood search: Principles and applications. European Journal of Operational Research **130**, 449–467 (2001)
75. Harris, S., Bueter, T., Tauritz, D.: A comparison of genetic programming variants for hyper-heuristics. In: Proceedings of the 2015 Annual Conference on Genetic Programming and Evolutionary Computation (GECCO'15), pp. 1043–1050 (2015)
76. Haspeslagh, S., Causmaecker, P.D., Schaerf, A., Stølevik, M.: The first international nurse rostering competition 2010. Annals of Operations Research **218**(1), 221–236 (2014)
77. Homberger, J., Gehring, H.: A two-phase hybrid metaheuristic for the vehicle routing problem with time windows. European Journal of Operational Research **162**(1), 220–238 (2005)
78. Hong, L., Drake, J.H., Woodward, J.R., Ozcan, E.: A hyper-heuristic approach to automated generation of mutation operators for evolutionary programming. Applied Soft Computing **62**, 162–175s (2018)
79. Hruska, F., Kubalik, J.: Selection hyper-heuristic using a portfolio of derivative heuristics. In: Proceedings of the Companion Publication of the 2015 Annual Conference on Genetic and Evolutionary Computation (GECCO'15), pp. 1401–1402 (2015)

80. Hsiao, P.C., Chiang, T.C., Fu, L.C.: A variable neighbourhood search-based hyper-heuristic for cross-domain optimization problems in CHeSC 2011 competition. http://www.asap.cs.nott.ac.uk/external/chesc2011/entries/hsiao-chesc.pdf (2011)
81. Hsiao, P.C., Chiang, T.C., Han, L.: A VNS-based hyper-heuristic with adaptive computational budget of local search. In: Proceedings of the WCCI 2012 World Congress on Computational Intelligence, pp. 1–8 (2012)
82. Hyde, M.: A genetic programming hyper-heuristic approach to automated packing. Ph.D. thesis, School of Computer Science, University of Nottingham (2010)
83. Ikegami, A., Niwa, A.: A subproblem-centric model and approach to the nurse scheduling problem. Mathematical Programming **97**(3), 517–541 (2003)
84. Jacobsen-Grocott, J., Mei, Y., Chen, G., Zhang, M.: Evolving heuristics for dynamic vehicle routing with time windows using genetic programming pp. 1948–1955 (2017)
85. Jia, Y., Cohen, M., Harman, M., Petke, J.: Learning combinatorial interaction test generaiton strategies using hyper-heuristics search. In: Proceedings of the 37th IEEE Conference on Software Engineering, pp. 540–550 (2015)
86. Jin, Y.: A comprehensive survey of fitness approximation in evolutionary computation. Soft Computing **9**, 3–12 (2005)
87. Jones, T.: Crossover, macromutation, and population-based search. In: Proceedings of the Sixth International Conference on Genetic Algorithms, pp. 73–80 (1995)
88. Keller, R., Poli, R.: Self-adaptive hyper-heuristic and greedy search. In: Proceedings of 2008 IEEE World Congress on Computational Intelligence (WCCI'08), pp. 3801–3801. IEEE (2008)
89. Kendall, G., Cowling, P.: Choice function and random hyperheuristics. In: Springer (ed.) Proceedings of the Fourth Asia-Pacific Conference on Simulated Evolution and Learning (SEAL), pp. 667–671 (2002)
90. Kendall, G., Hussin, N.: An investigation of a tabu-search-based hyper-heuristic for examination timetabling. In: S.P. G. Kendall E.K. Burke, M. Gendreau (eds.) Multidisciplinary Scheduling: Theory and Applications, pp. 309–328 (2005)
91. Kheiri, A., Keedwell, E.: Markov chain selection hyper-heuristic for the optimisation of constrained magic squares. In: UKCI 2015: UK Workshop on Computational Intelligence (2015)
92. Kheiri, A., Keedwell, E.: A sequence-based selection hyper-heuristic utilising a hidden markov model. In: Proceedings of 2015 Annual Conference on Genetic and Evolutionary Computation, pp. 417–424 (2015)
93. Kheiri, A., Özcan, E.: An iterated multi-stage selection hyper-heuristic. European Journal of Operational Research **250**, 77–90 (2016)
94. Kilby, P., Prosser, P., Shaw, P.: Dynamic VRPs: A study of scenarios. In: Report APES-06-1998, http://www.cs.strath.ac.uk/ apes/apereports.html. University of Strathclyde (1998)
95. Kiraz, B., Uyar, A.S., Ozcan, E.: An investigation of selection hyper-heuristics in dynamic environments. EvoApplications: Applications of Evolutionary Computations, Lecture Notes in Computer Science **6624**, 314–323 (2011)
96. Koza, J.: Genetic Programming: On the Programming of Computers by Means of Natural Selection, 1st edn. MIT (1992)
97. Laporte, G., Gendreau, M., Potvin, J., Semet, F.: Classical and modern heuristics for the vehicle routing problem. International Transactions in Operational Research **7**, 285–300 (2000)
98. Larose, M.: A hyper-heuristic for the CHeSC 2011. http://www.asap.cs.nott.ac.uk/external/chesc2011/entries/larose-chesc.pdf (2011)
99. Lassouaoui, M., Boughaci, D., Benhamou, B.: A hyper-heuristic method for MAX-SAT. In: Proceedings of the International Conference on Metaheuristics and Nature Inspired Computer (META'14), pp. 1–3 (2014)
100. Lehre, P., Özcan, E.: A runtime analysis of simple hyper-heuristics: To mix or not to mix operators. In: Proceedings of the Twelfth Workshop on Foundations of Genetic Algorithms, pp. 97–104 (2009)
101. Lenstra, J., Kan, A.: Complexity of vehicle routing and scheduling problems. Networks **11**(2), 221–227 (1981)

102. Li, D., Zhan, R., Zheng, D., Li, M., Kaku, I.: A hybrid evolutionary hyper-heuristic approach for intercell scheduling considering transportation capacity. IEEE Transactions on Automation Science and Engineering **12**(2), 1072–1089 (2016)
103. Li, J., Burke, E., Qu, R.: Integrating neural networks and logistic regression to underpin hyper-heuristic search. Knowledge-Based Systems **24**(2), 322–330 (2010)
104. Liu, Y., Mei, Y., Zhang, M., Zhang, Z.: Automated heuristic design using genetic programming hyper-heuristic for uncertain capacitated arc routing problem pp. 290–297 (2017)
105. López-Camacho, E., Terashima-Marin, H., Ross, P., Ochoa, G.: A unified hyper-heuristic framework for solving bin packing problems. Expert Systems with Applications **41**, 6876–6889 (2014)
106. Lourenco, H., Martin, O., Stutzle, T.: Handbook of Metaheuristics, *International Series in Operations Research and Management Science*, vol. 57, chap. Iterated Local Search, pp. 320–353. Springer (2003)
107. Lourenco, N., Pereira, F., Costa, E.: The importance of the learning conditions in hyper-heuristics. In: Proceedings of the 15th Annual Conference on Genetic and Evolutionary Computation, pp. 1525–1532 (2013)
108. Maashi, M., Kendall, G., Özcan, E.: Choice function based hyper-heuristics for multi-objective optimization. Applied Soft Computing **28**, 312–326 (2015)
109. Martello, S., Pisinger, D., Vig, D.: The three-dimensional bin packing problem. Operations Research **48**(2), 256–267 (2000)
110. McKay, R., Hoai, N., Whigham, P., Shan, Y., O'Neill, M.: Grammar-based genetic programming: A survey. Genetic Programming and Evolvable Machines **11**(3), 365–396 (2010)
111. Mei, Y., Zhang, M.: A comprehensive analysis on reusability of GP-evolved job shop dispatching rules. In: Proceedings of the 2016 IEEE Congress on Evolutionary Computation (CEC'16), pp. 3590–3597 (2016)
112. Meignan, D.: An evolutionary programming hyper-heuristic with co-evolution for chesc'11. http://www.asap.cs.nott.ac.uk/external/chesc2011/entries/meignan-chesc.pdf (2011)
113. Meignan, D., Koukam, A., Creput, J.: Coalition-based metaheuristic: a self-adaptive metaheuristic using reinforcement learning and mimetism. Journal of Heuristics **16**(6), 859–879 (2010)
114. Merlot, L., Boland, N., Hughes, B., Stuckey, P.: A hybrid algorithm for the examination timetabling problem. In: Burke, E. and Causmaecker, P. (eds.): Selected Papers from the 4th International Conference on the Practice and Theory of Automated Timetabling, pp. 207–231. Lecture Notes in Computer Science 2740 (2002)
115. Merz, P., Freisleben, B.: Fitness landscapes, memetic algorithms, and greedy operators for graph bipartitiioning. Evolutionary Computation **1**, 61–91 (2000)
116. Miranda, P., Prudencio, R., Pappa, G.: H3ad: A hybrid hyper-heuristic for algorithm design. Information Sciences **414**, 340–354 (2017)
117. Misir, M., Causmaecker, P.D., Berghe, G.V., Verbeeck, K.: An adaptive hyper-heuristic for chesc 2011. http://www.asap.cs.nott.ac.uk/external/chesc2011/entries/misir-chesc.pdf (2011)
118. Misir, M., Verbeeck, K., Causmaecker, P.D., Berghe, G.: Hyper-heuristics with a dynamic heuristic set for the home care scheduling problem. In: Proceedings of 2010 IEEE Congress on Evolutionary Computation (CEC'2010), p. 10.1109/CEC.2010.5586348 (2010)
119. Misir, M., Verbeeck, K., Causmaecker, P.D., Berghe, G.: An investigation on the generality level of selection hyper-heuristics under different empirical conditions. Applied Soft Computing **13**(7), 3335–3353 (2013)
120. Misir, M., Verbeeck, K., Causmaecker, P.D., Berghe, G.V.: A new hyper-heuristic as a general problem solver: An implementation in hyflex. Journal of Scheduling **16**, 291–311 (2013)
121. Misir, M., Wauters, T., Verbeeck, K., Berghe, G.: A hyper-heuristic with learning automata for the travelling tournament problem. In: Metaheuristics: Intelligent Decision Making, chap. 21, pp. 325–345. Springer (2012)
122. Mlejnek, J., Kubalik, J.: Evolutionary hyperheuristic for capacitated vehicle routing problem. In: The 15th Annual Conference on Genetic and Evolutionary Computation, pp. 219–220 (2013)

123. Mole, R., Jameson, S.: A sequential route-building algorithm employing a generalised savings criterion. Operational Research Quarterly **27**, 503–511 (1976)
124. Nguyen, S., Zhang, M., Johnston, M.: A genetic programming based hyper-heuristic approach for combinatorial optimization. In: Proceedings of the Genetic and Evolutionary Computation Conference (GECCO'15), pp. 1299–1306. ACM (2015)
125. Nyathi, T., Pillay, N.: Comparison of a genetic algorithm to grammatical evolution for automated design of genetic programming classification algorithms. Expert Systems with Applications **104**, 213–234 (2018)
126. Ochoa, G., Hyde, M., Curtois, T., Vazquez-Rodriguez, J.A., Walker, J., Gendreau, M., Kendall, G., McCollum, B., J.Parkes, A., Petrovi, S., Burke, E.K.: Hyflex: A benchmark framework for cross-domain heuristic search. In: Lecture Notes in Computer Science (EvoCOP 2012), vol. 7245, pp. 136–147. Springer (2012)
127. Ochoa, G., Qu, R., Burke, E.: Analyzing the landscape of a graph based hyper-heuristic for timetabling problems. In: The Genetic and Evolutionary Computation Conference (GECCO'09), pp. 341–348 (2009)
128. Ochoa, G., Veerapen, N.: Deconstructing the big valley search space hypothesis. In: Chicano F, Hu B, García-Sánchez P (ed.) Evolutionary Computation in Combinatorial Optimization: 16th European Conference (EvoCOP 2016), pp. 58–73 (2016)
129. O'Neill, M., Ryan, C.: Grammatical Evolution: Evolutionary Automatic Programming in an Arbitrary Language. Springer (2003)
130. Osogami, T., Imai, H.: Classification of various neighborhood operations for the nurse scheduling problem. In: Technical Report 135. The Institute of Statistical Mathematics (2000)
131. Özcan, E., Misir, M., Burke, E.: A self-organizing hyper-heuristic framework. In: Proceedings of the Multidisciplinary International Conference on Scheduling: Theory and Applications (MISTA 2009), pp. 784–787 (2009)
132. Özcan, E., Misir, M., Ochoa, G., Burke, E.: A reinforcement learning-great-deluge hyper-heuristic for examination timetabling. International Journal of Applied Metaheursitic Computing pp. 39–59 (2010)
133. Özcan, E., Parkes, A.: Policy matrix evolution for generation of heuristics. In: Proceedings of the 13th Annual Conference on Genetic and Evolutionary Computation, pp. 2011–2018 (2011)
134. Papadimitriou, C., Steiglitz, K.: Combinatorial Optimization: Algorithms and Complexity. Dover (1998)
135. Petrovic, S., Qu, R.: Cased-based reasoning as a heuristic selector in a hyper-heuristic for course timetabling problems. In: Knowledge-Based Intelligent Information Engineering Systems and Applied Technologies, Proceedings of KES'02, vol. 82, pp. 336–340 (2002)
136. Pillac, V., Gendreau, M., Guéret, C., Medaglia, A.: A review of dynamic vehicle routing problems. Networks **225**(1), 1–11 (2013)
137. Pillay, N.: Evolving hyper-heuristics for the uncapacitated examination timetabling problem. In: Proceedings of the Multidisciplinary International Conference on Scheduling, pp. 409–422 (2009)
138. Pillay, N.: Evolving heuristics for the school timetabling problem. In: Proceedings of the 2011 IEEE Conference on Intelligent Computing and Intelligent Systems (ICIS 2011), vol. 3, pp. 281–286 (2011)
139. Pillay, N.: Evolving hyper-heuristics for the uncapacitated examination timetabling problem. Journal of Operational Research Society **63**(47-58) (2012)
140. Pillay, N.: A study of evolutionary algorithm hyper-heuristics for the one-dimensional bin-packing problem. South African Computer Journal **48**, 31–40 (2012)
141. Pillay, N.: Evolving construction heuristics for the curriculum based university course timetabling problem. In: Proceedings of the IEEE Congress on Evolutionary Computation (CEC'16), pp. 4437–4443. IEEE (2016)
142. Pillay, N.: A review of hyper-heuristics for educational timetabling. Annals of Operations Research **239**(1), 3–38 (2016)

143. Pillay, N., Banzhaf, W.: A study of heuristic combinations for hyper heuristic systems for the uncapacitated examination timetabling problem. European Journal of Operational Research **197**, 482–491 (2009)
144. Pillay, N., Banzhaf, W.: An informed genetic algorithm for the examination timetabling problem. Applied Soft Computing **10**, 457–467 (2010)
145. Pillay, N., Beckedahl, D.: EvoHyp-a Java toolkit for evolutionary algorithm hyper-heuristics. In: Proceedings of the 2017 IEEE Congress on Evolutionary Computation, pp. 2707–2713s (2017)
146. Pillay, N., Ozcan, E.: Automated generation of constructive ordering heuristics for educational timetabling. Annals of Operations Research pp. https://doi.org/10.1007/s10,479–017–2625–x (2017)
147. Pillay, N., Qu, R.: Assessing hyper-heuristic performance. European Journal of Operational Research (under review) (2018)
148. Pillay, N., Rae, C.: A survey of hyper-heuristics for the nurse rostering problem pp. 115–122 (2012)
149. Pisinger, D., Ropke, S.: A general heuristic for vehicle routing problems. Computers & Operations Research **34**(8), 2403–2435 (2007)
150. Poli, R., Graff, M.: There is a free lunch for hyper-heuristics, genetic programming and computer scientists. In: European Conference on Genetic Programming (EuroGP 2009), pp. 195–207 (2009)
151. Qu, R., Burke, E.: Hybridisations within a graph based hyper-heuristic framework for university timetabling problems. Journal of Operational Research Society **60**, 1273–1285 (2009)
152. Qu, R., Burke, E., McCollum, B.: Adaptive automated construction of hybrid heuristics for exam timetabling and graph colouring problems. European Journal of Operational Research **198**(2), 392–404 (2009)
153. Qu, R., Burke, E., McCollum, B., Merlot, L., Lee, S.: A survey of search methodologies and automated system development for examination timetabling. Journal of Scheduling **12**(1), 55–89 (2009)
154. Qu, R., Pham, N., Bai, R., Kendall, G.: Hybridising heuristics within an estimation distribution algorithm for examination timetabling. Applied Intelligence **42**(4), 679–693 (2015)
155. Qu, R., Pillay, N.: A theoretical framework for hyper-heuristics. IEEE Transactions on Evolutionary Computation (under review) (2017)
156. Rae, C., Pillay, N.: Investigation into an evolutionary algorithm hyper-heuristic for the nurse rostering problem. In: Proceedings of the 10th International Conference on the Practice and Theory of Automated Timetabling, pp. 527–532 (2014)
157. Raghavjee, R., Pillay, N.: A genetic algorithm selection perturbative hyper-heuristic for solving the school timetabling problem. ORiON **31**(1), 39–60 (2015)
158. Reinelt, G.: Tsplib, a traveling salesman problem library. ORSA Journal on Computing **3**(4), 376–384 (1991)
159. Ritzinger, U., Puchinger, J., Hartl, R.: A survey on dynamic and stochastic vehicle routing problems. International Journal of Production Research **54**(1), 215–231 (2016)
160. Ross, P., Marin-Blazquez, J., Hart, E.: Hyper-heuristics applied to class and exam timetabling problems. In: Proceedings of the IEEE Congress of Evolutionary Computation CEC'04, pp. 1691–1698 (2004)
161. Ross, P., Marn-Blazquez, J., Schulenburg, S., Hart, E.: Learning a procedure that can solve hard bin-packing problems: A new GA-based approach to hyper-heuristics. In: Lecture Notes in Computer Science - GECCO 2003, vol. 2724, pp. 1295–1306. Springer (2003)
162. Ross, P., Schulenburg, S., Marin-Blazquez, J., Hart, E.: Hyper-heuristics: Learning to combine simple heuristics in bin-packing problems. In: Proceedings of the Genetic and Evolutionary Computation Conference, GECCO'02, pp. 942–948 (2002)
163. Ryser-Welch, P., Miller, J.F., Asta, S.: Generating human-readable algorithms for the travelling salesman problem using hyper-heuristics. In: Proceedings of the Companion Publication of the 2015 Annual Conference on Genetic and Evolutionary Computation, pp. 1067–1074. ACM (2015)

164. Sabar, N., Ayob, M., Kendall, G., Qu, R.: Grammatical evolution hyper-heuristic for combinatorial optimization problems. IEEE Transactions on Evolutionary Computation **17**(6), 840–861 (2013)
165. Sabar, N., Ayob, M., Qu, R., Kendall, G.: A graph colouring constructive hyper-heuristic for examination timetabling problems. Applied Intelligence **37**(1), 1–11 (2012)
166. Sabar, N., Zhang, X., Song, A.: A math-hyper-heuristic approach for large-scale vehicle routing problems with time windows, pp. 830–837 (2015)
167. Sabar, N.R., Ayob, M., Kendall, G., Qu, R.: Automatic design of a hyper-heuristic framework with gene expression programming for combinatorial optimization problems. IEEE Transactions on Evolutionary Computation **19**(3), 309–325 (2015)
168. Sabar, N.R., Kendall, G.: Population based Monte Carlo tree search hyper-heuristic. Information Sciences **314**, 225–239 (2015)
169. Sabar, N.R., Kendall, G., Qu, R.: A dynamic multi-armed bandit-gene expression programming hyper-heuristic for combinatorial optimization problems. IEEE Transactions on Cybernetics **45**(2), 217–228 (2015)
170. Saint-Guillain, M., Devill, Y., Solnon, C.: A multistage stochastic programming approach to the dynamic and stochastic VRPTW. In: International Conference on AI and OR Techniques in Constraint Programming for Combinatorial Optimization Problems, pp. 357–374. Springer (2015)
171. Scholl, A., Klein, R., Jurgens, C.: Bison: A fast hybrid procedure for exactly solving the one-dimensional bin packing problem. Computers and Operations Research **24**(7), 5–30 (1997)
172. Segredo, E., Lalla-Ruiz, E., Hart, E., Paechter, B., Voss, S.: Hybridization of evolutionary algorithms through hyper-heuristics for global continuous optimization. In: Proceedings of the International Conference on Learning and Intelligent Optimization (LION 2016), pp. 296–305 (2016)
173. Shahriar, A., Özcan, E., Curtois, T.: A tensor based hyper-heuristic for nurse rostering. Knowledge-Based Systems **98**(1), 185–199 (2016)
174. Sim, K., Hart, E.: Generating single and multiple cooperative heuristics for the one dimensional bin packing problem using a single node genetic programming island model. In: Proceedings of the 5th Annual Conference on Genetic and Evolutionary Computation(GECCO'13), pp. 1549–1556. ACM (2013)
175. Sim, K., Hart, E.: A combined generative and selective hyper-heuristic for the vehicle routing problem. In: Proceedings of the 21st Annual Conference on Genetic and Evolutionary Computation (GECCO'16), pp. 1093–1100 (2016)
176. Solomon, M.: Algorithms for the vehicle routing and scheduling problems with time window constraints. Operations Research **35**(2), 254–265 (1987)
177. Soria-Alcaraz, J., Ochoa, G., Sotelo-Figeroa, M., Burke, E.: A methodology for determining an effective subset of heuristics in selection hyper-heuristics. European Journal of Operational Research **260**(3), 972–983 (2017)
178. Sosa-Ascencio, A., Ochoa, G., Terashima-Marin, H., Conant-Pablos, S.: Grammar-based generation of variable-selection heuristics for constraint satisfaction problems. Genetic Programming and Evolvable Machines **17**(2), 119–144 (2015)
179. Swan, J., Causmaecker, P.D., Martin, S., Özcan, E.: A re-characterization of hyper-heuristics. In: L. Amodeo, E.G. Talbi, F. Yalaoui (eds.) Recent Developments of Metaheuristics, pp. 1–16. Springer (2016)
180. Swan, J., Woodward, J., Özcan, E., Kendall, G., Burke, E.: Searching the hyper-heuristic design space. Cognitive Computation **6**(1), 66–73 (2014)
181. Terashima-Marin, H., Ortiz-Bayliss, J., Ross, P., Valenzuela-Rendon, M.: Hyper-heuristics for the dynamic variable ordering in constraint satisfaction problem. In: Proceedings of the 10th Annual Conference on Genetic and Evolutionary Computation (GECCO'08), pp. 571–578. ACM (2008)
182. Terashima-Marín, H., Ross, P., López-Camacho, E., Valenzuela-Rendón, M.: Generalized hyper-heuristics for solving 2D regular and irregular packing problems. Annals of Operations Research **179**, 369–392 (2010)

183. Terashima-Marín, H., Ross, P., Valenzuela-Rendón, M.: Evolution of constraint satisfaction strategies in examination timetabling **1**, 635–642 (1999)
184. Terashima-Marin, H., Zarate, C., Ross, P., Valenzuela-Rendon, M.: A ga-based method to produce generalized hyper-heuristics for the 2D-regular cutting stock problem. In: Proceedings of the 8th Annual Conference on Genetic Programming and Evolutionary Algorithms, pp. 591–598. ACM (2006)
185. Toth, P., Vigo, D.: Models, relaxations and exact approaches for the capacitated vehicle routing problem. Discrete Applied Mathematics **123**(13), 487–512 (2002)
186. Toth, P., Vigo, D.: An overview of vehicle routing problems. In: The Vehicle Rrouting Problem, pp. 1–26 (2002)
187. Tyasnurita, R., Özcan, E., John, R.: Learning heuristic selection using a time delay neural network for open vehicle routing. In: 2017 IEEE Congress on Evolutionary Computation, pp. 1474–1481 (2017)
188. Valouxis, C., Housos, E.: Hybrid optimisation techniques for the workshift and rest assignment of nursing personnel. Artificial Intelligence in Medicine **20**, 155–175 (2000)
189. Vázquez-Rodríguez, J., Petrovic, S.: A new dispatching rule based genetic algorithm for the multi-objective job shop problem for the multi-objective job shop problem. Journal of Heuristics **16**, 771–793 (2010)
190. Walker, D.J., Keedwell, E.: Multi-objective optimisation with a sequence-based selection hyper-heuristic. In: Proceedings of the 2016 Companion Conference on Genetic and Evolutionary Computation, pp. 81–82 (2016)
191. Walker, J., Ochoa, G., Gendreau, M., Burke, E.: Vehicle routing and adaptive iterated local search within the hyflex hyper-heuristic framework, pp. 265–276 (2012)
192. Weinberger, E.: Correlated and uncorrelated fitness landscapes and how to tell the difference. Biological Cybernetics **63**, 325–336 (1990)
193. Weise, T., Devert, A., Tang, K.: A developmental solution to (dynamic) capacitated arc routing problems using genetic programming, pp. 831–838 (2012)
194. Whitley, D., Watson, J.: Complexity theory and the no free lunch theorem. In: Burke, E.K. and Kendall, G. (eds.) Search Methodologies: Introductory Tutorials in Optimization and Decision Support Techniques, Chapter. 11, pp. 317–339 (2005)

Index

N. Pillay, R. Qu, *Hyper-Heuristics: Theory and Applications*,
Natural Computing Series, https://doi.org/10.1007/978-3-319-96514-7

Printed in the United States
By Bookmasters